Metabotropic GPCRs: TGR5 and P2Y Receptors in Health and Diseases

Senthilkumar Rajagopal • Murugavel Ponnusamy

Metabotropic GPCRs: TGR5 and P2Y Receptors in Health and Diseases

Springer

Senthilkumar Rajagopal
Department of Biochemistry
Rayalaseema University
Kurnool, Andhra Pradesh, India

Murugavel Ponnusamy
Center for Developmental Cardiology
Qingdao University
Qingdao, Shandong, China

ISBN 978-981-13-1570-1 ISBN 978-981-13-1571-8 (eBook)
https://doi.org/10.1007/978-981-13-1571-8

Library of Congress Control Number: 2018957108

This Springer imprint is published by the registered company Springer Nature Singapore Pte Ltd.
The registered company address is: 152 Beach Road, #21-01/04 Gateway East, Singapore 189721, Singapore

Foreword A

G-protein-coupled receptors (GPCRs), known variously as heptahelical receptors or 7TM receptors, play important roles in detecting molecules outside the cell. As such, GPCRs have gained considerable scientific attention in the pathogenesis of diseases and drug discovery. As suggested, GPCRs act as a bridge between interior and exterior biochemical signaling and thus affect biological and pathological processes such as taste, vision, immunity, behaviour, neurological disorders, metabolic diseases and cancer. More than 35% of known drugs on the market, as well as pharmacological agents under development, act either directly or indirectly on GPCRs.

In recent years, the field of GPCR research has taken a decidedly multidisciplinary approach, and taking advantage of new techniques in structural, molecular and computational biology, we have seen continuous advancements in understanding the molecular signature, physiological role and pathological functions of each member in this large protein family of receptors.

Accordingly, this book extends our knowledge of the field by focusing on the metabotropic GPCRs, TGR5 and P2Y. The writers, Senthilkumar Rajagopal and Murugavel Ponnusamy, bring insight to these two ubiquitously expressed receptors that represent druggable targets for a wide range of disorders, including metabolic and cardiovascular diseases, in the context of molecular biology, experimental research, pharmacology and clinical diagnosis, which, taken together, highlight the most active areas of GPCR study today. The introductory chapter offers students and new researchers a comprehensive overview of GPCR biology. The following chapters explain the effects of TGR5 and P2Y receptors and their associated signaling pathways in the physiology and pathophysiology of various organs and systemic disorders in a compositional style that will have broad appeal across disciplines. The last chapter is a perspective that illustrates the advancements in drug discovery related to GPCR research and the chemical and pharmacological agents currently available to target these two receptors. This gives all readers a window through which to view and understand the expansion of this field of research, in general, and, more specifically, the future direction of these two receptors.

In sum, this work is certain to be a deskside companion and reference volume for those just starting their scientific careers and those who are experienced independent physician-scientists. I enthusiastically recommend this book to students, investigators and clinicians whose work extends into the field of metabotropic GPCR research.

Sakthivel Sadayappan
Professor of Internal Medicine,
Director of Heart Branch, Heart, Lung
and Vascular Institute,
Division of Cardiovascular Health and Disease
University of Cincinnati
Cincinnati, OH, USA

Foreword B

I am honoured to write the Foreword to this book *Metabotropic GPCRs: TGR5 and P2Y Receptors in Health and Diseases* written and edited by *Dr. Senthilkumar Rajagopal* and *Dr. Murugavel Ponnusamy*. Comprehending GPCR-associated physiology is a wonderfully resilient venture. As the research progresses, there are new and interesting findings that have the scope for translation for the benefit of mankind. The authors have the expertise and in-depth knowledge to systematically imbibe the complex topics of GPCR-associated physiology published in various journals and make it easier for a vast number of readers, including clinicians, academicians and beginners who have a passion for GPCR signaling. GPCRs are plasma membrane-associated receptors with seven transmembrane domains and hence popularly called 'serpentine family' receptors. Their mechanism of action is highly exciting but complex as they are implicated in a wide range of clinical and pathological conditions, and hence they are the attractive drug targets. This book has well-written four chapters with updated material to account for new findings in GPCR research and new exploring illustrations with an exciting bibliography. I am sure the robust effort of the authors in bringing this book to the public domain certainly enriches the scientific temperament of the reader.

Professor, Department of Biochemistry
University of Mysore
Mysuru, India

K. Kemparaju

Contents

About the Authors

Dr. Senthilkumar Rajagopal is an Assistant Professor at Rayalaseema University, Kurnool, AP, India, and obtained his Ph.D. in Biochemistry from Annamalai University, India. He did his postdoctoral research in Anesthesiology in the University of Virginia, Virginia, followed by Department of Cancer Immunology & AIDS, Dana-Farber Cancer Institute, Harvard University, MA, USA, and Department of Physiology and Biophysics, Virginia Commonwealth University, Virginia, USA. He is a recipient of Rameshwardasji Birla Smarak Kosh Endowment award from Mumbai Medical Trust, Mumbai, India, and Ramalingaswami Re-entry Fellowship from the Department of Biotechnology, Ministry of Science and Technology, Government of India. He has more than 10 years of research experience in various fields such as alcoholic hepatotoxicity, voltage-gated calcium channels, protein kinase C phosphorylation, ADAM family, smooth muscle physiology, G-protein-coupled receptor signaling pathways, and neurotransmitter transporters to central nervous system.

He has published nearly 48 research articles in various peer-reviewed international journals and 18 book chapters and presented research articles in a number of conferences proceedings and attended various workshops/seminars. He is on the editorial/reviewer board in many peer-reviewed international journals such as *Frontiers in Cellular Neuroscience*, *American Journal of Biochemistry and Molecular Biology*, and *Neuroscience*. He is a member of various prestigious scientific associations like the Indian Society for Atherosclerosis Research (ISAR), International Science Congress Association (ISCA), Society of Biological Chemistry (SBC), Society for Free Radical Research (SFRR) India, and Indian Society of Cell Biology (ISCB) and member of the American Association of Pharmaceutical Scientists (AAPS) and Science Advisory Board (SAB).

Dr. Murugavel Ponnusamy is presently working as Postdoctoral Research Associate at the Center for Developmental Cardiology, Institute for Translational Medicine, Qingdao University, China. Previously, he had worked as Postdoctoral Fellow at the Division of Renal Diseases, Department of Medicine, Rhode Island Hospital, Warren Alpert Medical School, Brown University, USA, under the supervision of Dr. Shougang Zhuang. His research interest includes molecular pharmacology, cellular signaling, tissue repair and regeneration, and experimental therapeutics. Currently, his research interest is focused on the influence of noncoding RNAs in cardiovascular disorders. His Ph.D. research was focused on elucidating the effect of diallyl tetrasulfide on cadmium-induced toxicity (an in vivo and in vitro study), which was carried at Annamalai University, India. Till now, he has published more than 40 research articles in various international peer-reviewed journals and contributed 5 chapters to different books. He is an active member of various societies including the Society of Biological Scientists of India (SOBSI), Indian Society of Cell Biology (ISCB), and Society of Biological Chemists (SBC) India.

Chapter 1
Overview of G-Protein Coupled Receptor

Abstract G protein coupled receptors (GPCRs) are the largest family of membrane proteins in the human genome, and substantial sources of targets for many industries. During the past two decades, the growth of GPCRs biology is very remarkable. The important findings of the GPCRs includes, first GPCRs cloning, and the sequencing of the human genome revealing the size of the GPCRs family and the number of orphan GPCRs. The N-terminus of the GPCRs protein present outside of the cell and C-terminus appears inside of the cell and having seven transmembrane-spanning segments, the main feature of GPCRs. The different sets of ligands such as proteins, small molecules, hormones, drugs and photons can bind to GPCRs astoundingly in their N-terminus and formed a pocket in the extracellular and transmembrane domains. Indeed, the numbers of studies have shown that GPCRs regulation and signaling is not easy to analyze than originally visualized, and includes signaling through G protein independent pathways. Here we have given a detailed mechanism of signal transduction through GPCR/G proteins and their function and structure.

Keywords G-protein · α-helices · Ligands · Signaling pathway · Transmembrane receptor

1.1 Introduction

A combination of large family of receptors protein is called as G protein–coupled receptors (GPCRs), and is also known as seven-(pass)-transmembrane domain receptors, 7TM receptors, heptahelical receptors, serpentine receptor, and G protein–linked receptors (GPLR). A molecules appears in outside of the cells detects these receptors and activate internal signal transduction pathways and, finally, cellular responses. The receptors bond with G proteins and they are called seven-transmembrane receptors because they pass through the cell membrane seven times (Hazell et al. 2012; Kobilka 2007; Venkatakrishnan et al. 2013). GPCRs are mostly expressed in eukaryotes, including yeast, choanoflagellates, and animals. The ligands are binds to these receptors and activate the receptors. These ligands are

S. Rajagopal, M. Ponnusamy, *Metabotropic GPCRs: TGR5 and P2Y Receptors in Health and Diseases*, https://doi.org/10.1007/978-981-13-1571-8_1

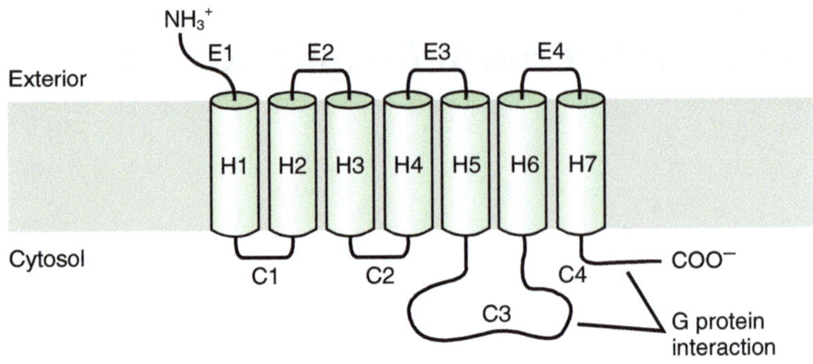

Fig. 1.1 Schematic diagram of the general structure of G protein-coupled receptors. (**a**) The receptors are present in similar orientation in the membrane. (**b**) Comprise seven transmembrane alpha helical regions (H1–H7). C) Four extracellular segments (E1–E4). (**d**) Four cytosolic segments (C1–C4). (**e**) The carboxy-terminal segment (C4), the C3 loops, and, in some receptors, also the C2 loop are involved in interactions with a coupled trimeric G protein

light-sensitive compounds, odors, pheromones, hormones, and neurotransmitters, and vary in size from small molecules to peptides to large proteins. Most of the modern medicinal drugs are targeting these GPCRs which are involved in many diseases (Ji et al. 1998; Knoflach et al. 2001; Tuteja 2009) (Fig. 1.1).

1.2 Classification

Among human genome sequence, an accurate size of the GPCR super family is unknown, moreover, 800 different human genes of GPCR family is covered by only 4% of the entire protein-coding genome, which is identified from genome sequence analysis. A number of classification have been proposed for the GPCR super family and it can be classified into three main classes (A, B and C) with no observed shared sequence homology between classes (Geng-Ming et al. 2017; Hilger et al. 2018; Wu et al. 2017; Zhang et al. 2015). Class A contains 80% of the GPCR genes and 50% of the receptors encode to olfactory receptors and remaining receptors classified as orphan receptors. The structure and function of the GPCR is similar since the homology between the classes is not known so far. The very large rhodopsin a group has been further subdivided into 19 subgroups (A1-A19). The recent studies have shown that there is an alternative classification system called GRAFS (Glutamate, Rhodopsin, *Adhesion*, Frizzled/Taste2, and Secretin) (Digby et al. 2006; Latorraca et al. 2017; Penela et al. 2003; Tan et al. 2004). Based on the classical A-F system, GPCRs can be grouped into 6 classes based on sequence homology and functional similarity:

- Class A (or 1) (Rhodopsin-like)
- Class B (or 2) (Secretin receptor family)

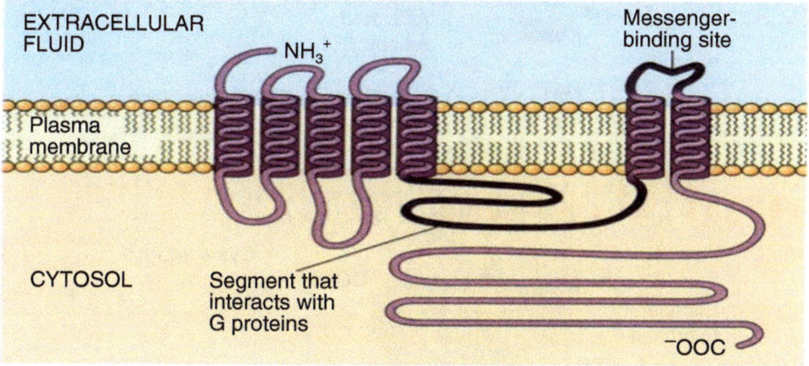

Fig. 1.2 G-Proteins: G protein-linked receptors have an extracellular N-terminus and a cytosolic C-terminus separated by 7 transmembrane α-helices connected by peptide loops. One of the extracellular segments has a unique messenger binding site. The cytosolic loop between the 5th and 6th α-helices specifically binds a particular G protein. Activated G proteins bind to enzymes or other proteins and alter the target protein's activity

- Class C (or 3) (Metabotropic glutamate/pheromone)
- Class D (or 4) (Fungal mating pheromone receptors)
- Class E (or 5) (Cyclic AMP receptors)
- Class F (or 6) (Frizzled/Smoothened)

Recent studies have revealed that the human genome encodes approximately 750 GPCRs, based on available DNA sequence, most of the hormones, growth factors, endogenous ligands can detect about 350 of these GPCRs and approximately, 150 of the GPCRs found in the human genome have unknown functions. Based on GPCR's pseudo amino acid composition approach, GPCR's classified into different ways, using their amino acid sequence alone (Ellis 2004; Gaulton and Attwood 2003; Lapinsh et al. 2002; Zhen-Ling et al. 2010) (Fig. 1.2).

The GPCRs are acting through two main signal transduction pathways such as cAMP and phosphatidylinositol signal pathways. GPCRs play a significant role in different signaling pathways via switching many extracellular signals and transduce them to heterotrimeric G proteins, which further transduce these signals intracellular to appropriate downstream effectors. G Proteins activated by the GPCRs after conformational change of themself. Gα-GTP of GTP is split and Gα becomes inactive (Gα-GDP), which leads to its re-association with the Gβ/Gγ dimer to form the inactive heterotrimeric complex after the signal propagation. The GPCRs can send the signal through G protein independent pathway also. The dissociation of G protein's α subunit, from the β and γ subunits can directly target the functional proteins depending on the α subunit type ($G_{\alpha s}$, $G_{\alpha i/o}$, $G_{\alpha q/11}$, $G_{\alpha 12/13}$) and to further affect intracellular signaling proteins. GPCRs also regulate cell cycle progression (Cabrera-Vera et al. 2003; Elling et al. 1997; Hauser et al. 2017; Kobilka 2007; Rosenbaum et al. 2009; Schiöth and Fredriksson 2005; Schonenbach et al. 2014) (Fig. 1.3).

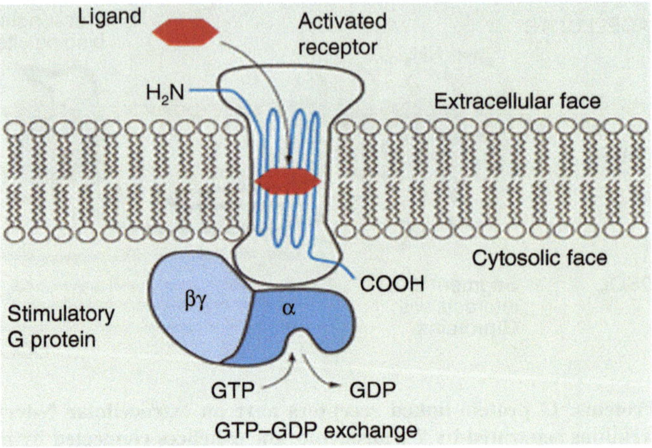

Fig. 1.3 Signal send-off. Extracellular region of GPCR'S triggers changes in transmembrane region of protein after ligand binding. This causes the release of guanosine diphosphate (GDP) and the uptake of guanosine triphosphate (GTP) from the G protein, spurring activation of predefined signalling pathways

1.3 Physiological Roles

Most of the physiological processes in humans are regulated by GPCRs. Few examples of their physiological roles include: (a) Visual Sense (b) Gustatory Sense (c) Sense of Smell (d) Regulation of Behavioural and mood (e) Regulation of immune system activity and inflammation (f) Autonomic nervous system transmission (g) Cell density sensing.

In Photoisomerization reactions, the visual sense receptors use the opsins to translate electromagnetic radiation into cellular signals. For instance, Rhodopsin uses the conversion of 11-cis-retinal to all-trans-retinal for this purpose. The taste receptors in taste cells secrets gustducin after exposure to bitter and sweet-tasting substances. GPCRs present in brain controls the secretion of several different neurotransmitters such as serotonin, dopamine, GABA and glutamate which is modulates the behavioural and mood function of the human and animals (Avau and Depoortere 2016; Jaggupilli et al. 2016; Kolakowski 1994; Xue et al. 2008).

Cells of the immune system triggers intracellular communication between them after chemokine receptors bind ligands and receptors such as histamine receptors bind inflammatory mediators and engage target cell types in the inflammatory response. GPCRs are also involved in immune-modulation and directly involved in suppression of TLR-induced immune responses from T cells (Allen et al. 2007; David and Krzysztof 2009; Kakinuma and Hwang 2006). The sympathetic and parasympathetic nervous systems are responsible for control of many automatic functions of the body such as blood pressure, heart rate and digestive processes. These two nervous systems are regulated by GPCRs pathways. A novel GPCR role in regulating cell density sensing. (a) Homeostasis modulation (e.g., water balance). (b) Involved in growth and metastasis of some types of tumours (Bunnett and Cottrell 2010; Krishnan and Schiöth 2015; Yunhong and Amantha 2015).

1.4 GPCR Signaling – Activation and Regulation

GPCRs guide a multiple range of intracellular signaling cascades in response to growth factors, ions, hormones, neurotransmitters, etc. Thus, they have vital roles in many cellular events, including proliferation, differentiation, cell-cell communication and chemotaxis. As a cell surface receptor, GPCRs are activated by a diverse set of ligands that transduce extracellular signal and amplify intracellular signaling through second messengers to generate a variety of cellular responses. The signaling ability of GPCRs is affected by their localization in a cell membrane. The proper assembly of extracellular N-terminus is greatly important for the ligand binding and receptor is trafficking. In the GPCR super family, a highly conserved hydrophobic transmembrane domain, composed of seven helices structure traversing membrane (7TM), connects the extracellular and intracellular signaling. The binding of ligands in the transmembrane and/or extracellular domain trigger the conformational changes in GPCR, which initiates intracellular events (Katritch et al. 2012, 2013). The great diversity of GPCR family members are seen in the N-terminus, the intracellular loop of TM5 and TM6 and C-terminus, which are responsible for the ligand specificity, differential interaction and fine tuning of intracellular signaling. In fact, the cytoplasmic domains are also well conserved in terms of shape and size, which facilitates the specific binding and interaction of a common set of signaling partners (Katritch et al. 2012, 2013). The biophysical studies with X-ray crystallography revealed that ligand binding regions are divided into two pockets (designated as main or major binding pocket and minor binding pocket) by the angles of transmembrane helices. The conformational changes occur in the GPCR depends on the binding affinity of ligands to these pockets and this conformational change is a key for the activation and propagation of specific cytoplasmic signaling. The major binding pocket is located between extracellular segments of TM3-7 that alters G-protein binding without affecting arrestin pathway. The minor ligand-binding minor pocket is located between the extracellular ends of TM1, 2, 3 and 7, which affects arrestin signaling (Rosenkilde et al. 2010; Sojka et al. 2017). Some ligands have the capability to bind with both major and minor pockets and this binding negatively influences both G-protein and arrestin signaling (Sojka et al. 2017) (Fig. 1.4).

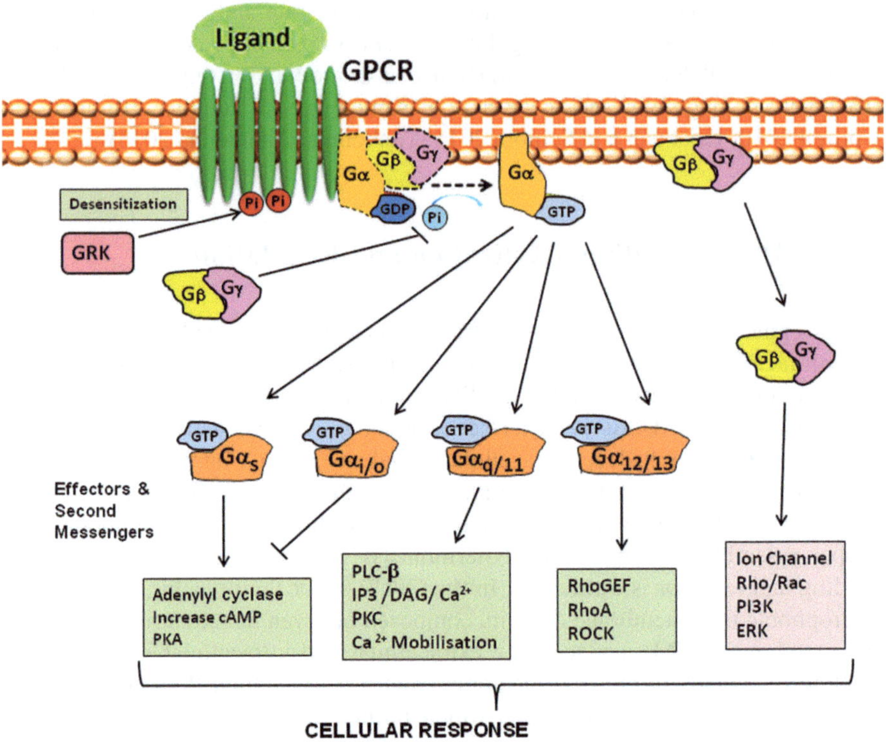

Fig. 1.4 General scheme of GPCR signaling

1.5 Mode of Activation

Several novel modes of GPCRs activation are identified in the biological system and that depicts the complexity of GPCR signaling. In addition to the conventional GPCR-G protein activation mode, biased activation, intracellular activation, dimerization/oligomerization dependent activation, transactivation and biphasic activation are most frequently occurring during the initiation of GPCR signaling (Fig. 1.5). In the classical pathway of GPCRs, the activated GPCR interacts with intracellular heteromeric G-proteins, composed of three subunits Gα, Gβ and Gγ, to pass the extracellular message to the downstream signaling (Katritch et al. 2012). The ligand bound GPCR acts as a GDP/GTP exchanging factor, which promotes binding of GTP to Gα by displacement of GDP. The binding of GTP causes conformational changes in Gα, which leads to its dissociation from a tightly bound Gβγ complex. these dissociated subunits interacts with a variety of effectors molecules in order to activate downstream signaling. Gα subunit targets adenylyl cyclase, phospholipase C (PLC), cyclic GMP phosphodiesterase and RhoGTPase nucleotide exchange factors (RhoGEF). On the other hand, Gβγ-dimeric complex is primarily involved in the regulation of GPCRs by recruiting GRKs and inactivation of Gα in order to

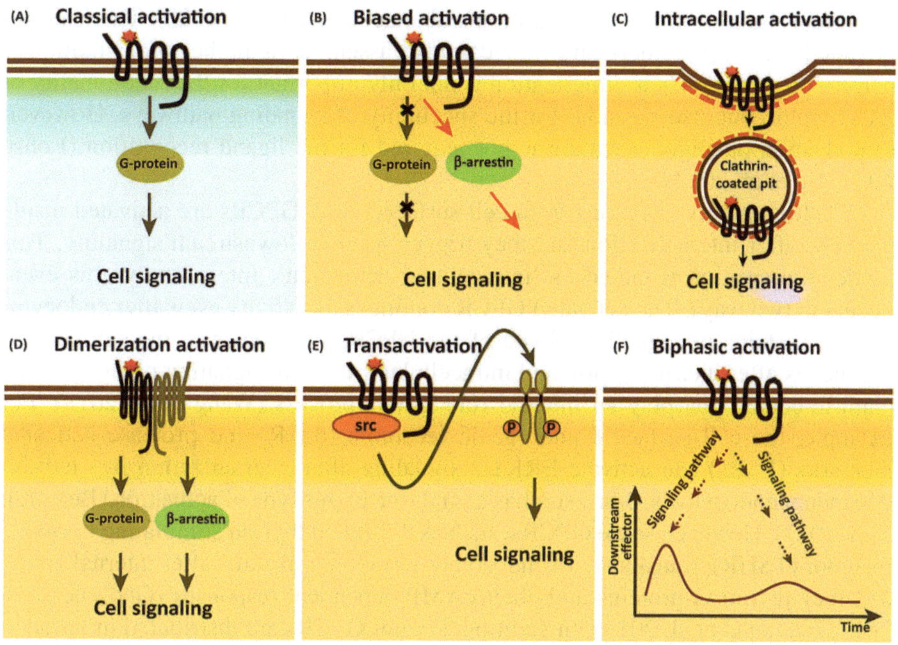

Fig. 1.5 Different mode of GPCR action

terminate GPCR signaling. It also regulates many signaling pathways including voltage-dependent Ca^{2+} channels, Src, PLC, PI3K, PyK2, and MAPK (Hilger et al. 2018; Wang et al. 2018). Apart from this, β-arrestins, a multifunctional adaptor protein, can initiate G-protein independent GPCR signal activation. They activate several signaling molecules such as c-Src, extracellular regulated kinase (ERK), Janus activated kinase (JNK) and small GTPase RhoA by forming a complex with them. It can also translocate to the nucleus and promotes transcription of various genes by acting as a scaffold for transcription factors such as p300, CREB and hypoxia inducible factor 1α (Cianfrocca et al. 2016; Ma and Pei 2007).

Some GPCR ligands activate either G-protein or β-arrestin; this event is called as 'biased activation'. G-proteins dependent signals turn on effectors through modulation of the levels of second messengers (cAMP, IP3, etc) by activating enzymes such as PLC or adenylyl cyclase. In contrast, β-arrestin activates ERK1/2 and other kinases for accelerating the GPCR signal. These two pathways are temporally distinct and they have different physiological and pathological effects. In general, β-arrestin opposes G-proteins activity and signaling. The classical example for this type of activation is, type I angiotensin receptor (AT1R) dependent activation of G-protein and β-arrestin signaling. The activation of G-protein by AT1R promotes vasoconstriction and increases blood pressure, while the activation of β-arrestin signaling protects against toxicants by stimulating cell survival pathways (Ahn et al. 2009; Pupo et al. 2016). In some instance, these two pathways have synergetic effects. AT1R-dependent chemotaxis requires integration of β-arrestin and

G-protein signals for the cell migration and attachment (Defea 2008). The majority of GPCRs exist as dimers and they can also assemble as higher order homo- or hetero- oligomers. Their dimerization is not only important for the functionality of the receptor, but also essential for the specificity of signaling pathways. However, dimerization or oligomerization is not required for the ligand recognition (Lohse 2010; Ng et al. 2012).

GPCRs not only functions from cell surface, many GPCRs are activated inside the cells after internalization and they trigger specific downstream signaling. This mode of activation is called as 'intracellular activation'. Interestingly, this event occurs in two ways. The activated GPCR continues its activity even after endocytosis and sustains its signaling. Some of the GPCRs translocate to the sub cellular organelles after internalization and intracellularly activate signaling pathways. The clathrin-coated pits play an integral role in this process (Wang et al. 2018). For example, the cell surface β-androgenic receptors (β-AR) and protease activated receptors (PARs) can activate ERK1/2 signaling after internalization by clathrin-dependent endocytosis. β-arrestin has a vital role in this type of activation (Beautrait et al. 2017). However, some GPCRs, such as β-AR and thyroid stimulating hormone receptor (TSHR), translocate to endosomes or Golgi apparatus after internalization and they recruit G-proteins and elicit cAMP dependent responses (Godbole et al. 2017; Irannejad et al. 2013). In addition, several GPCRs are distributed in intracellular organelles such as mitochondria, lysosomes, melanosomes, nuclear membrane, etc. where they are involved in internalization-independent intracellular signaling activation (Wang et al. 2018). For example, metabotropic glutamate receptor 5 (mGluR5) residing in inner nuclear membrane, coupling to $G_{q/11}$, contributes to the regulation of Ca^{2+} changes in the nucleoplasm (Sergin et al. 2017).

GPCRs have the capability to transactivate receptor tyrosine kinases (RTKs) and protein serine/threonine kinase receptors. These two families of receptors often functions as partners and they mutually activate each other. Many RTKs including epidermal growth factor receptor (EGFR), transforming growth factor β receptor (TFβR) and toll like receptor (TLRs) are transactivated by a variety of GPCRs (Chaplin et al. 2017). On the other hand, the receptors of insulin and platelet derived growth factor (PDGF) have the capability to transactivate GPCR family members including chemokine receptor 4 (CXCR4), β2-adregenic receptor (β2AR) and sphingosine-1-phosphate receptor (S1P1) (Akekawatchai et al. 2005; Cattaneo et al. 2005; Kamato et al. 2015). Some of the GPCRs couple with RTKs to integrate intracellular signals. This integration increases the efficiency and also sustains the effectors signaling (Pyne and Pyne 2011). The components of GPCR signaling such as G-proteins, β-arrestin and regulator of G protein signaling (RGS) act as either downstream or upstream of RTKs. In addition, many cytoplasmic signaling molecules such as c-Src, Src homology and collagen (Shc), Son of sevenless (SoS), Ras, Yes associated protein (YAP)/TAZ and MAPK are integral components of GPCR-RTK complex (Chaplin et al. 2017; Yu et al. 2012). Another unique mode of GPCR signaling is 'biphasic response' during which GPCR transiently activates one intracellular signal within few minutes of ligand binding (early phase) and return to basal level. The second wave of activity (late phase) results in sustained activation of same signaling pathway or different pathways. In this mode of activation, GPCRs exhibit two distinct biological consequences due to variation in duration and intensity of GPCRs on their downstream effectors. For instance, β3 adrenergic receptor (β3-AR) induces smooth muscle cell proliferation by biphasic activation of ERK1/2 signaling. In the early phase (after 3 minutes of receptor activation), β3-AR stimulates Gs-PKA-Src axis to activate ERK1/2, but it activates ERK1/2 in Gi-Src-PI3K pathway dependent manner during the late phase (after 8 hours of stimulation) (Hadi et al. 2013). Similarly, GPCR dependent activation of YAP-TAZ signaling contributes to both growth stimulation and growth suppression in intestinal epithelial cells. The GPCR agonists such as AngII and vasopressin rapidly increase YAP phopshorylation and nuclear exclusion within 15 minutes after stimulation through $G_{\alpha q/11}$/PKC/PKD signaling pathway, but this translocation is transient and GPCR/PKD signal promotes YAP re-localization to nucleus by Rho-dependent activation in the late phase. This event sustains transcriptional activity of YAP and promotes cell proliferation in intestinal epithelial cells (Wang et al. 2016).

1.6 Regulation of GPCR Signaling

Most of the cells have naturally developed a way to limit and control GPCR activity. The receptor desensitization is a universal mechanism involved in the silencing of GPCR signal, which is classically orchestrated by G-protein coupled receptor kinases (GRKs) and arrestins mediated phosphorylation of cytoplasmic loop of activated GPCRs. GRKs are members of regulators of G protein signaling (RGS), which are vital components required for the suppression of GPCR signaling. RGS proteins act as GTPase-accelerating proteins (GAPs), which promote G protein deactivation by hydrolyzing G-protein bound GTP and resulting in termination of GPCR downstream signal. The RGS family can influence signals initiated by both α and βγ subunits of G proteins by acting on $G_{\alpha i}$ and Gαq. The activity of RGS determines the intensity and duration of cellular responses initiated by GPCR signals. RGS proteins also contribute to protein translation, translocation of signaling molecules and protein-protein interaction by serving as a scaffold molecule. Thus, they are not just committed to inhibit GPCR signaling, but also modulate and integrate many signaling pathways (Sjogren et al. 2010; Woodard et al. 2015). RGS family consists of more than 36 members and all these contain a conserved RGS domain (also known as RGS box), composed of 120 amino acids, which binds specifically to activated Gα. Based on the similarities and alignment of RGS domain amino acid sequence, they are divided into four subfamilies (A/RZ, B/R4, C/R7, and D/R12) (Louwette et al. 2012; Soundararajan et al. 2008). In addition, several proteins consist of non-functional RGS homology domain with weaker or without GAP activity, which are called as RGS-like proteins. They are classified into E/RA, F/GEF, G/GRK, H/SNX and I/D-AKAP subfamilies. A typical examples for these classes of proteins are Axin, GPCR kinase 2 (GRK2), p115RhoGEF and LARG (Louwette et al. 2012) (Fig. 1.6).

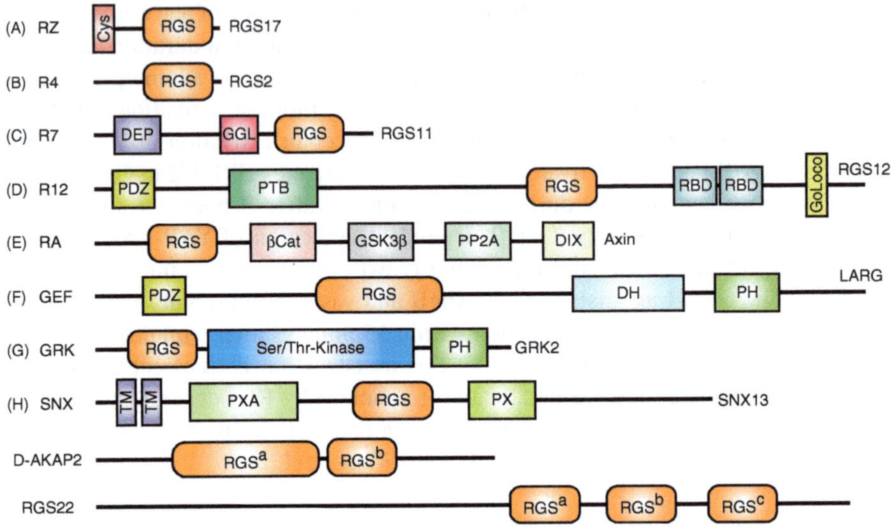

Fig. 1.6 RGS classification based on structure and homology of RGS domain

RGS proteins selectively target certain GPCRs irrespective of the couple G proteins and regulate the functions of those GPCRs. The variation in splicing (alternative splicing) is primarily responsible for the selectivity, specificity and spatio-temporal expression. A best example for this mechanism is the tissue distribution, expression pattern and functions of alternatively spliced RGS9 (RGS9-1 and RGS9-2). RGS9-1 is more specifically expressed in rod and cone photoreceptor cells of retina, in which RGS9-1 selectively interacts with Gαt and regulates photo transduction. It also blocks cGMP production by directly binding to guanylyl cyclase and inhibiting its activity. RGS9-2 is undetectable in retina. In contrast, RGS9-2 is predominantly expressed in the brain tissue and it specifically regulates dopamine D2 and μ-opioid receptor by targeting Gi and Gq members such as Gαi, Gαo, Gαz, Gαq and Gα11 proteins (Xie and Palmer 2007). The genomic studies revealed that RGS are often arranged adjacent to G-protein related components in gene clusters. The tight genomic coupling of G-protein components and RGS facilitates the functional regulation of these genes at transcription and protein levels (Sierra et al. 2002). The crystallographic studies found that specific sequence within the RGS domains and Gα subunits determines the interaction and strength of affinity these molecules. In the case of RGS4-Gαi1 interaction, Asn^{128} residue of RGS domain specifically binds with Gln^{204}, Ser^{206}, and Glu^{207} of Gαi1 subunits. In fact, Asp^{117}, Ser^{118}, and Arg^{121} residues in the RGS domain form a unique binding pocket for Thr^{182} of Gαi1. But GAIP/RGS19 has differential selectivity and affinity to Gαi1 subunits although the three dimensional structure is very similar to RGS4. This is mainly due to 40% variation in residues of RGS domain in these two RGS and binding affinity of Gαi to ser^{156} instead of Asn^{128} in RGS domain (Xie and Palmer 2007). In addition, C-terminal domain of RGS protein also takes part in the recognition of

G-protein by serving as an affinity adaptor. The length of C-terminal changes the strength of affinity with its target G-protein (Martemyanov et al. 2003). Thus, the amino acid configuration of RGS proteins, in particularly RGS domain and C-terminal region, determines the binding specificity, strength and time of RGS and G-protein interaction.

1.7 GPCR in Development and Physiology

Many GPCRs contribute to the activation and regulation of various stages of the developmental processes including spermatogenesis, oocyte development and maturation, embryogenesis and organogenesis in mammalian system. Several members of rhodopsin-like GPCR, adhesion-like GPCR and secretin-like GPCR are important contributors in every stage of the gestation, from fertilization to birth, for the normal development of the offspring. Here, we illustrate some of the well-defined GPCRs and their role in the developmental process. The expression and heteromerization of GPCRs is vital for the regulation of reproductive hormone secretion, proper growth and maturation of follicles and oocytes in female reproductive system (Satake et al. 2013). Several members of rhodopsin-like GPCR, adhesion-like GPCR and secretin-like GPCR are important contributors in the embryonic development. In the early developmental stage, the functions of GPCRTre1 (in drosophila) and CXCR4 (in vertebrates) is crucial for the migration of primordial germ cells. In addition, adhesion-like GPCR, secretin-like GPCR and olfactory GPCRs are important partners in the proliferation of germ line progenitors as well as gamete morphogenesis (Kunwar et al. 2003; Molyneaux et al. 2003). CXCR4 is a unique G-coupled chemokine receptor, which plays fundamental role in the development of the hematopoietic, nervous and cardiovascular system during embryogenesis. The CXCL12 (ligand) and its receptors (CXCR4 and CXCR7) is the well conserved chemokine signaling axis in vertebrates, which activate both G-protein dependent and β-arrestin dependent pathways. CXCL12 signaling guides the migration of cells, which is a fundamental process in the dynamic environment of the developing embryo (Wang and Knaut 2014).

The quality of oocytes and its fertilization relies on the proper maturation of oocytes, which occurs by controlled meiosis process in order to avoid ploidy of gametes. A high level of cAMP promotes meiotic arrest and activate maturation promoting factor, composed of cyclin-dependent kinase 1 (CDK1) and cyclin B. Thus, the cAMP level determines the fate and quality of oocytes. GPR30, also known as GPER1, play a vital role in the maintenance of mitotic arrest and oocyte maturation in lower vertebrates such as fish by elevating cAMP level through stimulation of Gαs-adenylyl cyclase. GPR126 is an adhesion type of GPCR that is highly conserved in the vertebrates. The expression of GPR126 is required for proper development of cardiovascular system and lack of GPR126 results in mid-gestation lethality due to cardiac failure. In fact, the variation in the locus of GPR126 is important determinant of tallness in the human (Waller-Evans et al. 2010). In mammals, the arrest of oocytes in meiotic prophase from the time its development to ovulation, is crucial for the proper growth of oocyte before maturation. GPR3 protects the oocyte from early maturation and it is important for the female fertility. The oocytes with deficiency of GPR3 have reduced capacity for the development of embryo (Ledent et al. 2005). In addition to GPR3, GPR6, GPR12 and GPR185 (GPRx) are important mediators of meiotic arrest and they sustain the period of time of meiosis before maturation by increasing intracellular level of cAMP and/or cGMP (Rios-Cardona et al. 2008; Yang et al. 2012). GPR18 is a highly expressed non-olfactory GPCR in human spermatozoa. The N-arachidonoylglycine (NAGly) mediated activation of GPR18-PLCγ stimulates physiological reorganization of actin filaments sperm, which is crucial for its penetration into the oocyte and fertilization of egg (Flegel et al. 2016). Similarly, human testicular odorant receptor (hOR17-4) stimulates PKA dependent consecutive activation and nuclear translocation of β-arrestin2 in mature spermatozoa, which promotes fertilization by favorably triggering the expression of gene expression in spermatozoa required for the sperm-egg communication (Neuhaus et al. 2006). Egg activation is an initial step of embryogenesis, after fusion of sperm with egg, results in Ca^{2+} mobilization, expression and recruitment of maternal mRNA and cell cycle resumption. The heteromeric G proteins such as Gαs, Gαq and Gβγ are integral components of integral components of Ca^{2+} dynamics during fertilization by acting as upstream activators of cytoplasmic Ca^{2+} release (Voronina and Wessel 2004).

GPER1 is a distinct receptor from nuclear estrogen receptor that exhibits a strong spatiotemporal expression pattern during embryogenesis and it is mainly localized to brain, heart, neuromasts and craniofacial regions of early embryo. The estrogen dependent activation of GPER1 plays central role in the regulation of thyroid hormone (T3) level and heart rates in embryonic stage (Romano et al. 2017). The regulators of GPCRs such as GRK2 and GRK5 have fundamental role in cardiac morphology outcomes at developmental stage. The defective function or loss of these two GRKs causes many developmental abnormalities including dilated artria and ventricular hyperplasia (Philipp et al. 2014). Many GPCRs play vital role in the left-right patterning in vertebrate embryogenesis. GPR161, an orphan receptor, is highly expressed in primary cilia that promotes limb formation and patterning as well as proper skeletal morphogenesis by suppressing sonic hedgehog (Shh) signaling pathway by increasing Gli, which is a transcription suppressor of shh (Hwang and Mukhopadhyay 2015). In addition, GPR161 dependent activation of Gαs-cAMP axis and regulation of Shh-signaling is critical for the ventral patterning of the vertebrate neural tube (Hwang and Mukhopadhyay 2015). Molecular studies found that GPR161 regulates retinoic acid and classical Wnt pathways, and this regulation is indispensible for the neurulation and eye lens development (Matteson et al. 2008). Any defect or lack of GPR161 results in mid-gestation lethality in vertebrates by causing developmental and patterning defect in neural tubes, endochondral bones of forearms, bone collar and trebecular bone (Hwang and Mukhopadhyay 2015; Hwang et al. 2018). In zebra fish, the expression level of GPR161 in the lateral plate mesoderm determines of the heart looping and left-right axis patterning

(Leung et al. 2008). The lysophosphatidic acid (LPA) receptor 3 (LpaR3), a member of rhodopsin-like GPCR subfamily, also actively contribute to the left-right asymmetries of organs, in particularly cardiac looping, in zebra fish embryo through Wnt-signaling dependent regulation of Ca^{2+} mobilization (Lai et al. 2012). A study in *Xenopus* levis found that GPR84 also play central role in the development of retina, cornea and lens (Perry et al. 2010). The adhesion class GPCR, GPR126, controls outgrowth of canal projections and maintains semicircular canal duct structure through activating cAMP dependent pathway (Geng et al. 2013). The building of organs and their shaping rely on epithelial morphogenesis, which provides a base for the tissue compartmentalization and establishment of whole body plan in the early embryonic stage. It is also vital during gastrulation for the organization of the embryonic germ layers. In lower organisms like drosophila, Mist (a drosophila GPCR) play essential during differential developmental process including gastrulation, wing and leg morphogenesis by guiding folded gastrulation (Fog) and RhoGEF2 is signaling. GPRK2 tunes Fog signaling and it regulates cell morphogenesis and movements (Fuse et al. 2013; Manning et al. 2013). Interestingly, the GPCRs related GPR4, Flop1/2; regulate head formation in *Xenopus* by controlling the Wnt/β-catenin pathway (Miyagi et al. 2015).

1.8 Future Perspective

The physiological signaling of GPCRs is an intricate process and it remains largely unknown, but study of GPCR is quite interesting and it activates particular signaling cascade at different levels in the vary cells. Moreover, increasing evidences suggest that the complexity of the signaling mechanisms activated by GPCRs, the regulation of different proteins appears in cellular organization with which they interact, provide new challenges to our basic understanding of the pharmacology of this important class of cell surface receptors. In this chapter, we reviewed the characterization and physiological regulation of GPCR.

References

Ahn, S., Kim, J., Hara, M. R., Ren, X. R., & Lefkowitz, R. J. (2009). {beta}-Arrestin-2 mediates anti-apoptotic signaling through regulation of BAD phosphorylation. *Journal of Biological Chemistry, 284*, 8855–8865.

Akekawatchai, C., Holland, J. D., Kochetkova, M., Wallace, J. C., & McColl, S. R. (2005). Transactivation of CXCR4 by the insulin-like growth factor-1 receptor (IGF-1R) in human MDA-MB-231 breast cancer epithelial cells. *Journal of Biological Chemistry, 280*, 39701–39708.

Allen, S., Crown, S., & Handel, T. (2007). Chemokine: receptor structure, interactions, and antagonism. *Annual Review of Immunology, 25*, 787–820.

Avau, B., & Depoortere, I. (2016). The bitter truth about bitter taste receptors: beyond sensing bitter in the oral cavity. *Acta Physiology (Oxford), 216*, 407–420.

Beautrait, A., Paradis, J. S., Zimmerman, B., Giubilaro, J., Nikolajev, L., Armando, S., Kobayashi, H., Yamani, L., Namkung, Y., Heydenreich, F. M., Khoury, E., Audet, M., Roux, P. P., Veprintsev, D. B., Laporte, S. A., & Bouvier, M. (2017). A new inhibitor of the beta-arrestin/AP2 endocytic complex reveals interplay between GPCR internalization and signalling. *Nature Communications, 8*, 15054.

Bunnett, N., & Cottrell, G. (2010). Trafficking and signaling of G protein-coupled receptors in the nervous system: implications for disease and therapy. *CNS Neurological Disorders and Drug Targets, 9*, 539–556.

Cabrera-Vera, T., Vanhauwe, J., Thomas, T., Medkova, M., Preininger, A., Mazzoni, M., & Hamm, H. (2003). Insights into G protein structure, function, and regulation. *Endocrinology Review, 24*, 765–781.

Cattaneo, E., Zuccato, C., & Tartari, M. (2005). Normal huntingtin function: an alternative approach to Huntington's disease. *Nature Review Neuroscience, 6*, 919–930.

Chaplin, R., Thach, L., Hollenberg, M. D., Cao, Y., Little, P. J., & Kamato, D. (2017). Insights into cellular signalling by G protein coupled receptor transactivation of cell surface protein kinase receptors. *Journal of Cell Communication and Signalling, 11*, 117–125.

Cianfrocca, R., Tocci, P., Rosano, L., Caprara, V., Sestito, R., Di Castro, V., & Bagnato, A. (2016). Nuclear beta-arrestin1 is a critical cofactor of hypoxia-inducible factor-1alpha signaling in endothelin-1-induced ovarian tumor progression. *Oncotarget, 7*, 17790–17804.

David, T., & Krzysztof, P. (2009). Chemokine receptors and other GPCRs. *Current Opinion in HIV and AIDS, 4*, 88–95.

Defea, K. (2008). Beta-arrestins and heterotrimeric G-proteins: collaborators and competitors in signal transduction. *British Journal of Pharmacology, 153*(Suppl 1), S298–S309.

Digby, G., Lober, R., Sethi, P., & Lambert, N. (2006). Some G protein heterotrimers physically dissociate in living cells. *Proceedings of the National Academy of Sciences USA, 103*, 17789–17794.

Elling, C., Thirstrup, K., Nielsen, S., Hjorth, S., & Schwartz, T. (1997). Metal-ion sites as structural and functional probes of helix-helix interactions in 7TM receptors. *Annual New York Academic of Sciences, 814*, 142–151.

Ellis, C. (2004). The state of GPCR research in 2004. *Nature Review: Drug Discovery, 3*, 577–626.

Flegel, C., Vogel, F., Hofreuter, A., Wojcik, S., Schoeder, C., Kiec-Kononowicz, K., Brockmeyer, N. H., Muller, C. E., Becker, C., Altmuller, J., Hatt, H., & Gisselmann, G. (2016). Characterization of non-olfactory GPCRs in human sperm with a focus on GPR18. *Science Report, 6*, 32255.

Fuse, N., Yu, F., & Hirose, S. (2013). Gprk2 adjusts Fog signaling to organize cell movements in Drosophila gastrulation. *Development, 140*, 4246–4255.

Gaulton, A., & Attwood, T. (2003). Bioinformatics approaches for the classification of G-protein-coupled receptors. *Current Opinion in Pharmacology, 3*, 114–120.

Geng, F. S., Abbas, L., Baxendale, S., Holdsworth, C. J., Swanson, A. G., Slanchev, K., Hammerschmidt, M., Topczewski, J., & Whitfield, T. T. (2013). Semicircular canal morphogenesis in the zebrafish inner ear requires the function of gpr126 (lauscher), an adhesion class G protein-coupled receptor gene. *Development, 140*, 4362–4374.

Geng-Ming, H., Te-Lun, M., & Chi-Ming, C. (2017). Visualizing the GPCR Network: Classification and evolution. *Scientific Reports, 7*, 15495.

Godbole, A., Lyga, S., Lohse, M. J., & Calebiro, D. (2017). Internalized TSH receptors en route to the TGN induce local Gs-protein signaling and gene transcription. *Nature Communications, 8*, 443.

Hadi, T., Barrichon, M., Mourtialon, P., Wendremaire, M., Garrido, C., Sagot, P., Bardou, M., & Lirussi, F. (2013). Biphasic Erk1/2 activation sequentially involving Gs and Gi signaling is required in beta3-adrenergic receptor-induced primary smooth muscle cell proliferation. *Biochimica Biophysica Acta, 1833*, 1041–1051.

Hauser, A., Attwood, M., Rask-Andersen, M., Schiöth, H., & Gloriam, D. (2017). Trends in GPCR drug discovery: new agents, targets and indications. *Nature Reviews Drug Discovery, 16*, 829–842.

Hazell, G., Hindmarch, C., Pope, G., Roper, J., Lightman, S., Murphy, D., O'Carroll, A., & Lolait, S. (2012). G protein-coupled receptors in the hypothalamic paraventricular and supraoptic nuclei—Serpentine gateways to neuroendocrine homeostasis. *Frontiers in Neuroendocrinology, 33*, 45–66.

Hilger, D., Masureel, M., & Kobilka, B. (2018). Structure and dynamics of GPCR signaling complexes. *Nature: Structural and Molecular Biology, 25*, 4–12.

Hwang, S. H., & Mukhopadhyay, S. (2015). G-protein-coupled receptors and localized signaling in the primary cilium during ventral neural tube patterning. *Birth Defects Research Clinical Molecular Teratology, 103*, 12–19.

Hwang, S. H., White, K. A., Somatilaka, B. N., Shelton, J. M., Richardson, J. A., & Mukhopadhyay, S. (2018). The G protein-coupled receptor Gpr161 regulates forelimb formation, limb patterning and skeletal morphogenesis in a primary cilium-dependent manner. *Development, 145*.

Irannejad, R., Tomshine, J. C., Tomshine, J. R., Chevalier, M., Mahoney, J. P., Steyaert, J., Rasmussen, S. G., Sunahara, R. K., El-Samad, H., Huang, B., & von Zastrow, M. (2013). Conformational biosensors reveal GPCR signalling from endosomes. *Nature, 495*, 534–538.

Jaggupilli, A., Howard, R., Upadhyaya, J., Bhullar, R., & Chelikani, P. (2016). Bitter taste receptors: Novel insights into the biochemistry and pharmacology. *International Journal of Biochemistry and Cellular Biology, 77*, 184–196.

Ji, T., Grossmann, M., & Ji, I. (1998). G protein-coupled receptors. I. Diversity of receptor-ligand interactions. *Journal of Biological Chemistry, 273*, 17299–17302.

Kakinuma, T., & Hwang, S. (2006). Chemokines, chemokine receptors, and cancer metastasis. *Journal of Leukocyte Biology, 79*, 639–651.

Kamato, D., Rostam, M. A., Bernard, R., Piva, T. J., Mantri, N., Guidone, D., Zheng, W., Osman, N., & Little, P. J. (2015). The expansion of GPCR transactivation-dependent signalling to include serine/threonine kinase receptors represents a new cell signalling frontier. *Cellular and Molecular Life Sciences, 72*, 799–808.

Katritch, V., Cherezov, V., & Stevens, R. C. (2012). Diversity and modularity of G protein-coupled receptor structures. *Trends in Pharmacological Sciences, 33*, 17–27.

Katritch, V., Cherezov, V., & Stevens, R. C. (2013). Structure-function of the G protein-coupled receptor superfamily. *Annual Review of Pharmacology and Toxicology, 53*, 531–556.

Knoflach, F., Mutel, V., Jolidon, S., Kew, J., Malherbe, P., Vieira, E., Wichmann, J., & Kemp, J. (2001). Positive allosteric modulators of metabotropic glutamate 1 receptor: characterization, mechanism of action, and binding site. *Proceeding of Natural Acadmics of Sciences USA, 98*, 13402–13407.

Kobilka, B. (2007). G Protein coupled receptor structure and activation. *Biochemical and Biophysical Research Communication, 1768*, 794–807.

Kolakowski, L. (1994). GCRDb: a G-protein-coupled receptor database. *Receptors and Channels, 2*, 1–7.

Krishnan, A., & Schiöth, H. (2015). The role of G protein-coupled receptors in the early evolution of neurotransmission and the nervous system. *Journal of Experimental Biology, 218*, 562–571.

Kunwar, P. S., Starz-Gaiano, M., Bainton, R. J., Heberlein, U., & Lehmann, R. (2003). Tre1, a G protein-coupled receptor, directs transepithelial migration of Drosophila germ cells. *PLoS Biology, 1*, E80.

Lai, S. L., Yao, W. L., Tsao, K. C., Houben, A. J., Albers, H. M., Ovaa, H., Moolenaar, W. H., & Lee, S. J. (2012). Autotaxin/Lpar3 signaling regulates Kupffer's vesicle formation and left-right asymmetry in zebrafish. *Development, 139*, 4439–4448.

Lapinsh, M., Gutcaits, A., Prusis, P., Post, C., Lundstedt, T., & Wikberg, J. (2002). Classification of G-protein coupled receptors by alignment-independent extraction of principal chemical properties of primary amino acid sequences. *Protein Science, 11*, 795–805.

Latorraca, N., Venkatakrishnan, A., & Dror, R. (2017). GPCR dynamics: structures in motion. *Chemistry Review, 117*, 139–155.

Ledent, C., Demeestere, I., Blum, D., Petermans, J., Hamalainen, T., Smits, G., & Vassart, G. (2005). Premature ovarian aging in mice deficient for Gpr3. *Proceedings of the National Academy of Sciences, USA, 102*, 8922–8926.

Leung, T., Humbert, J. E., Stauffer, A. M., Giger, K. E., Chen, H., Tsai, H. J., Wang, C., Mirshahi, T., & Robishaw, J. D. (2008). The orphan G protein-coupled receptor 161 is required for left-right patterning. *Developmental Biology, 323*, 31–40.

Lohse, M. J. (2010). Dimerization in GPCR mobility and signaling. *Current Opinion in Pharmacology, 10*, 53–58.

Louwette, S., Van Geet, C., & Freson, K. (2012). Regulators of G protein signaling: role in hematopoiesis, megakaryopoiesis and platelet function. *Journal of Thrombosis and Haemost, 10*, 2215–2222.

Ma, L., & Pei, G. (2007). Beta-arrestin signaling and regulation of transcription. *Journal of Cell Science, 120*, 213–218.

Manning, A. J., Peters, K. A., Peifer, M., & Rogers, S. L. (2013). Regulation of epithelial morphogenesis by the G protein-coupled receptor mist and its ligand fog. *Science Signaling, 6*, ra98.

Martemyanov, K. A., Hopp, J. A., & Arshavsky, V. Y. (2003). Specificity of G protein-RGS protein recognition is regulated by affinity adapters. *Neuron, 38*, 857–862.

Matteson, P. G., Desai, J., Korstanje, R., Lazar, G., Borsuk, T. E., Rollins, J., Kadambi, S., Joseph, J., Rahman, T., Wink, J., Benayed, R., Paigen, B., & Millonig, J. H. (2008). The orphan G protein-coupled receptor, Gpr161, encodes the vacuolated lens locus and controls neurulation and lens development. *Proceedings of the National Academy of Sciences, USA, 105*, 2088–2093.

Miyagi, A., Negishi, T., Yamamoto, T. S., & Ueno, N. (2015). G protein-coupled receptors Flop1 and Flop2 inhibit Wnt/beta-catenin signaling and are essential for head formation in Xenopus. *Developmental Biology, 407*, 131–144.

Molyneaux, K. A., Zinszner, H., Kunwar, P. S., Schaible, K., Stebler, J., Sunshine, M. J., O'Brien, W., Raz, E., Littman, D., Wylie, C., & Lehmann, R. (2003). The chemokine SDF1/CXCL12 and its receptor CXCR4 regulate mouse germ cell migration and survival. *Development, 130*, 4279–4286.

Neuhaus, E. M., Mashukova, A., Barbour, J., Wolters, D., & Hatt, H. (2006). Novel function of beta-arrestin2 in the nucleus of mature spermatozoa. *Journal of Cell Science, 119*, 3047–3056.

Ng, S. Y., Lee, L. T., & Chow, B. K. (2012). Receptor oligomerization: from early evidence to current understanding in class B GPCRs. *Frontiers in Endocrinology (Lausanne), 3*, 175.

Penela, P., Ribas, C., & Mayor, F. (2003). Mechanisms of regulation of the expression and function of G protein-coupled receptor kinases. *Cellular Signalling, 15*, 973–981.

Perry, K. J., Johnson, V. R., Malloch, E. L., Fukui, L., Wever, J., Thomas, A. G., Hamilton, P. W., & Henry, J. J. (2010). The G-protein-coupled receptor, GPR84, is important for eye development in *Xenopus* laevis. *Development Dynasty, 239*, 3024–3037.

Philipp, M., Berger, I. M., Just, S., & Caron, M. G. (2014). Overlapping and opposing functions of G protein-coupled receptor kinase 2 (GRK2) and GRK5 during heart development. *Journal of Biological Chemistry, 289*, 26119–26130.

Pupo, A. S., Duarte, D. A., Lima, V., Teixeira, L. B., Parreiras, E. S. L. T., & Costa-Neto, C. M. (2016). Recent updates on GPCR biased agonism. *Pharmacological Research, 112*, 49–57.

Pyne, N. J., & Pyne, S. (2011). Receptor tyrosine kinase-G-protein-coupled receptor signalling platforms: out of the shadow? *Trends in Pharmacological Science, 32*, 443–450.

Rios-Cardona, D., Ricardo-Gonzalez, R. R., Chawla, A., & Ferrell, J. E., Jr. (2008). A role for GPRx, a novel GPR3/6/12-related G-protein coupled receptor, in the maintenance of meiotic arrest in *Xenopus* laevis oocytes. *Developmental Biology, 317*, 380–388.

Romano, S. N., Edwards, H. E., Souder, J. P., Ryan, K. J., Cui, X., & Gorelick, D. A. (2017). G protein-coupled estrogen receptor regulates embryonic heart rate in zebrafish. *PLoS Genetics, 13*, e1007069.

Rosenbaum, D., Rasmussen, S., & Kobilka, B. (2009). The structure and function of G-protein-coupled receptors. *Nature, 459*, 356–363.

Rosenkilde, M. M., Benned-Jensen, T., Frimurer, T. M., & Schwartz, T. W. (2010). The minor binding pocket: a major player in 7TM receptor activation. *Trends in Pharmacological Science, 31*, 567–574.

Satake, H., Matsubara, S., Aoyama, M., Kawada, T., & Sakai, T. (2013). GPCR Heterodimerization in the Reproductive System: Functional Regulation and Implication for Biodiversity. *Frontiers in Endocrinology (Lausanne), 4*, 100.

Schiöth, H., & Fredriksson, R. (2005). The GRAFS classification system of G-protein coupled receptors in comparative perspective. *General Comparative and Endocrinology, 142*, 94–101.

Schonenbach, N., Hussain, S., & O'Malley, M. (2014). Structure and function of G protein-coupled receptor oligomers: implications for drug discovery. *Nanomedicine and Nanomaterials, 7*, 408–427.

Sergin, I., Jong, Y. I., Harmon, S. K., Kumar, V., & O'Malley, K. L. (2017). Sequences within the C Terminus of the Metabotropic Glutamate Receptor 5 (mGluR5) Are Responsible for Inner Nuclear Membrane Localization. *Journal of Biological Chemistry, 292*, 3637–3655.

Sierra, D. A., Gilbert, D. J., Householder, D., Grishin, N. V., Yu, K., Ukidwe, P., Barker, S. A., He, W., Wensel, T. G., Otero, G., Brown, G., Copeland, N. G., Jenkins, N. A., & Wilkie, T. M. (2002). Evolution of the regulators of G-protein signaling multigene family in mouse and human. *Genomics, 79*, 177–185.

Sjogren, B., Blazer, L. L., & Neubig, R. R. (2010). Regulators of G protein signaling proteins as targets for drug discovery. *Progress in Molecular Biology and Translational Science, 91*, 81–119.

Sojka, A. C., Brennan, K. M., Maizels, E. T., & Young, C. D. (2017). The Science Behind G Protein-Coupled Receptors (GPCRs) and Their Accurate Visual Representation in Scientific Research. *Journal of Biocommunication, 41*, 32–44.

Soundararajan, M., Willard, F. S., Kimple, A. J., Turnbull, A. P., Ball, L. J., Schoch, G. A., Gileadi, C., Fedorov, O. Y., Dowler, E. F., Higman, V. A., Hutsell, S. Q., Sundstrom, M., Doyle, D. A., & Siderovski, D. P. (2008). Structural diversity in the RGS domain and its interaction with heterotrimeric G protein alpha-subunits. *Proceedings of the National Academy of Sciences, USA, 105*, 6457–6462.

Tan, C., Brady, A., Nickols, H., Wang, Q., & Limbird, L. (2004). Membrane trafficking of G protein-coupled receptors. *Annual Review of Pharmacology and Toxicology, 44*, 559–609.

Tuteja, N. (2009). Signaling through G protein coupled receptors. *Plant Signaling and Behavior, 4*, 942–947.

Venkatakrishnan, A., Deupi, X., Lebon, G., Tate, C., Schertler, G., & Babu, M. (2013). Molecular signatures of G-protein-coupled receptors. *Nature, 494*, 185–194.

Voronina, E., & Wessel, G. M. (2004). βγ subunits of heterotrimeric G-proteins contribute to Ca^{2+} release at fertilization in the sea urchin. *Journal of Cell Science, 117*, 5995–6005.

Waller-Evans, H., Promel, S., Langenhan, T., Dixon, J., Zahn, D., Colledge, W. H., Doran, J., Carlton, M. B., Davies, B., Aparicio, S. A., Grosse, J., & Russ, A. P. (2010). The orphan adhesion-GPCR GPR126 is required for embryonic development in the mouse. *PLoS One, 5*, e14047.

Wang, J., & Knaut, H. (2014). Chemokine signaling in development and disease. *Development, 141*, 4199–4205.

Wang, J., Sinnett-Smith, J., Stevens, J. V., Young, S. H., & Rozengurt, E. (2016). Biphasic Regulation of Yes-associated Protein (YAP) cellular localization, phosphorylation, and activity by G protein-coupled receptor agonists in intestinal epithelial cells: a novel role for Protein Kinase D (PKD). *Journal of Biological Chemistry, 291*, 17988–18005.

Wang, W., Qiao, Y., & Li, Z. (2018). New insights into modes of GPCR activation. *Trends in Pharmacological Science, 39*, 367–386.

Woodard, G. E., Jardin, I., Berna-Erro, A., Salido, G. M., & Rosado, J. A. (2015). Regulators of G-protein-signaling proteins: negative modulators of G-protein-coupled receptor signaling. *Internal Review of Cell and Molecular Biology, 317*, 97–183.

Wu, F., Song, G., de Graaf, C., & Stevens, R. (2017). Structure and function of peptide-binding G protein-coupled receptors. *Journal of Molecular Biology, 429*, 2726–2745.

Xie, G. X., & Palmer, P. P. (2007). How regulators of G protein signaling achieve selective regulation. *Journal of Molecular Biology, 366*, 349–365.

Xue, C., Yen-Ping, H., & Heitman, J. (2008). Magnificent seven: roles of G protein-coupled receptors in extracellular sensing in fungi. *FEMS Microbiology Reviews, 32*, 1010–1032.

Yang, C. R., Wei, Y., Qi, S. T., Chen, L., Zhang, Q. H., Ma, J. Y., Luo, Y. B., Wang, Y. P., Hou, Y., Schatten, H., Liu, Z. H., & Sun, Q. Y. (2012). The G protein coupled receptor 3 is involved in cAMP and cGMP signaling and maintenance of meiotic arrest in porcine oocytes. *PLoS One, 7*, e38807.

Yu, F. X., Zhao, B., Panupinthu, N., Jewell, J. L., Lian, I., Wang, L. H., Zhao, J., Yuan, H., Tumaneng, K., Li, H., Fu, X. D., Mills, G. B., & Guan, K. L. (2012). Regulation of the Hippo-YAP pathway by G-protein-coupled receptor signaling. *Cell, 150*, 780–791.

Yunhong, H., & Amantha, T. (2015). Regulation of neuronal communication by G protein-coupled receptors. *FEBS Letters, 589*, 1607–1619.

Zhang, D., Zhao, Q., & Wu, B. (2015). Structural studies of G protein-coupled receptors. *Molecular and Cell, 38*, 836–842.

Zhen-Ling, P., Jian-Yi, Y., & Chen, X. (2010). An improved classification of G-protein-coupled receptors using sequence-derived features. *BMC Bioinformatics, 11*, 420.

Chapter 2
TGR5 Receptor

Abstract Bile acid activated TGR5 is a G protein-coupled bile acid receptor 1 (GPBAR 1), which is also known as membrane type receptor (M-BAR). The single TGR5 exon gene encodes a protein of 330 amino acids and predicated seven transmembrane domain structures. The most of the tissues in humans such as endocrine glands, adipocytes, muscles, gall bladder, immune organs, spinal cord, and the enteric nervous system express TGR5 receptors. Whereas, liver and intestine have high levels expressions of nuclear receptor farenosid X-receptor (FXR). The primary and secondary bile acids act as agonist for this receptor and various steroid hormones as well. The mode of TGR5 activation varies depends on the tissue and it activates different signaling cascades depending on the cell type. The plasma membrane receptor TGR5 and nuclear receptor FXR can modulate endocrine function of bile acid. This chapter summarizes the basic information of TGR5 and FXR and their physiology and pathological functions.

Keywords Bile acids · Contraction and relaxation · Nuclear receptor · Plasma membrane receptor · Smooth muscle physiology

2.1 Introduction

Takeda G-Protein Receptor 5 (TGR5), also known as GPBAR1, M-BAR, or GPCR19, is the plasma membrane bound G protein coupled bile acid receptor 1 (GPBAR1), which is activated by bile acids. This receptor is localized in several tissues and it is highly expressed in lung, spleen, gallbladder, placenta brown adipose tissue (BAT) and intestine. Macrophages/monocytes most abundantly express TGR5 among the different type of cells (Maruyama et al. 2002). The membrane bound GPCR, TGR5 was discovered in 2002 (Maruyama et al. 2002). It was classified as the founder member of the bile acid receptor subclass of GPCRs (Foord et al. 2005). TGR5 gene locates on the position 2q35 in human chormosome and its open reading frame (ORF) has 993 base pairs encoding 330 amino acids. TGR5 mRNA is highly detectable in several organs such as small intestine, stomach, liver, lung-placenta, and spleen (Keitel et al. 2010; Tiwari and Maiti 2009). TGR5 is highly

S. Rajagopal, M. Ponnusamy, *Metabotropic GPCRs: TGR5 and P2Y Receptors in Health and Diseases*, https://doi.org/10.1007/978-981-13-1571-8_2

expressed in the epithelium of gallbladder and it enhances gallbladder filling with bile (Hua et al. 2016).

TGR5 is a member of the rhodopsin-like super family of GPCR that transmits signals through Gs protein. The endogenous ligands such as lithocholic acid (LCA) and taurolithocholic acid (TLCA) activate TGR5 at EC_{50} of 600 and 300 nM, respectively. TGR5 also activated by bile acids such as cholic acid (CA), deoxycholic acid (DCA), and chenodeoxycholic acid (CDCA) at higher micro molar concentrations. The bile acids activated TGR5 induces cAMP production (Maruyama et al. 2002). As a membrane receptor, the activation of TGR5 by ligands leads to its internalisation into the cytoplasm (Kawamata et al. 2003). TGR5 plays important roles in several cell signalling pathways including nuclear factor κB (NF-κB) (Chen et al. 2011), AKT (Kida et al. 2013), and extracellular signal regulated kinases (ERK) (Masyuk et al. 2013). Its endogenous and synthetic agonists could be potential therapeutic agents for the treatment of metabolic, inflammation and digestive disorders (Fig. 2.1) (Broeders et al. 2015; Kumar et al. 2012).

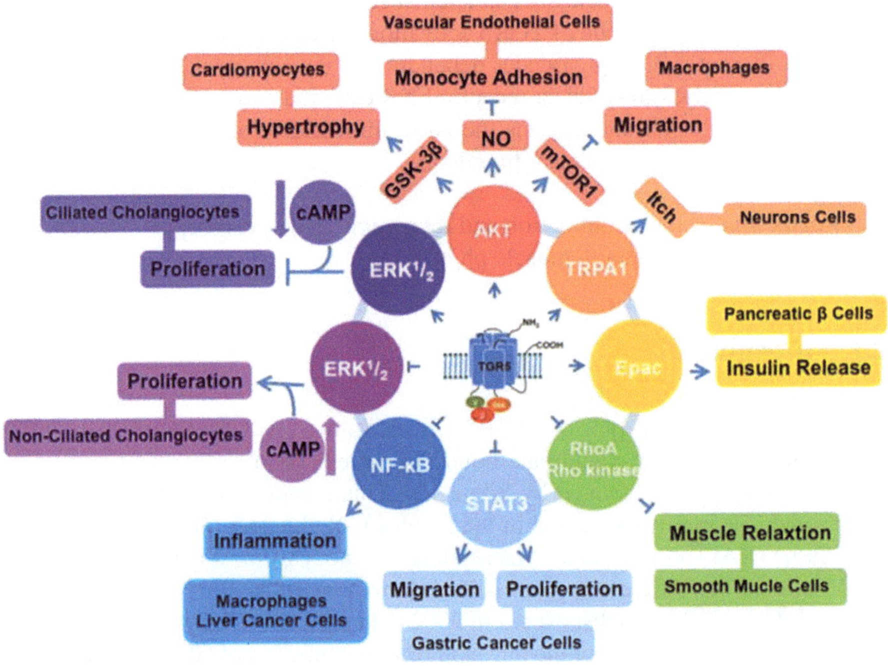

Fig. 2.1 TGR5 regulates different cell signalling pathways. TGR5 activates AKT (Kida et al. 2013; Perino et al. 2014), TRPA1 (Lieu et al. 2014) and Epac (Kumar et al. 2012) pathways. TGR5 inhibits NF-κB (Pols et al. 2011; Wang et al. 2011), STAT3 (Guo et al. 2015), and RhoA/Rho kinase (Rajagopal et al. 2013) pathways, while it has opposite functions in ERK1/2 pathway depending on the cell type. TGR5 activates ERK1/2 in ciliated cholangiocytes, (Masyuk et al. 2013) but activated TGR5 inhibits ERK1/2 in non-ciliated cholangiocytes (Masyuk et al. 2013)

2.2 TGR5 Signalling

Bile acids are formed in liver and secreted into duodenum in response to food. They play an important role in lipid homeostasis. Cholesterol is the precursor of bile acids. The major portion of bile acids are reabsorbed in the distal part of the ileum and colon and transported to the liver enterohepatic circulation (Kumar et al. 2012; Rajagopal et al. 2013). The bacterial enzymes of the intestine convert the primary bile acids such as CA and CDCA into their respective secondary bile acids such as DCA and LCA (Bala et al. 2011; Kumar et al. 2012). In addition to their classic function in lipid digestion, bile acids act as signaling molecules in bile acid synthesis and secretion, glucose and lipid metabolism, immune response, and cell proliferation and differentiation (Houten et al. 2006; Maruyama et al. 2002; McMillin et al. 2015; Per-Arne et al. 2013). In enteroendocrine cells, the activation of TGR5 increases the release of GLP-1, which maintains homeostasis of blood glucose by suppressing glucagon release and promoting glucose-induced insulin secretion, delaying gastric emptying, promoting satiety as well as increasing glucose disposal in the peripheral tissues (Katsuma et al. 2005).

TGR5 can also be activated by secondary bile acids including LCA and DCA, which are synthesized in the colon by gut bacteria. This type of TGR5 activation induces cAMP/PKA signaling to stimulate energy metabolism in BAT, relaxing and refilling of the gallbladder, and secretion of GLP-1 from intestinal endocrine L cells (Kumar et al. 2012). TGR5 could also be involved in the regulation of the alternative bile acid synthesis pathway by regulating CYP7B1, which is a key sexually dimorphic and male-predominant gene in the alternative bile acid synthesis pathway (Hiroki et al. 2016). The pro-inflammatory cytokine production and cholangiocyte proliferation inhibited by the activation of TGR5 in macrophages and Kupffer cells (Fig. 2.2) (Reich et al. 2016).

2.3 TGR5 and Immunometabolism

The expression and activation of TGR5 in monocytes and macrophages contributes to the regulation of inflammatory effects (Pols 2014).

2.3.1 Metabolic Inflammation

The proportions of metabolic diseases such as obesity and type 2 diabetes are steadily increasing due to change in life style including increased intake of calorie, reduced physical activity, combined with genetic predisposition. Inflammation is a key contributor in the development and progression of metabolic diseases and various mechanisms are involved in pathophysiology of these diseases. A low grade

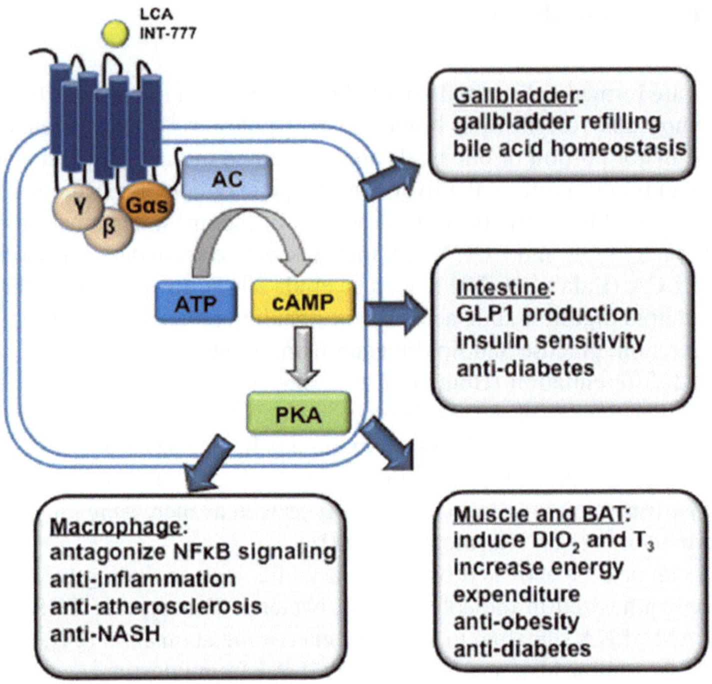

Fig. 2.2 Differential activation of TGR5 receptor in different organs

chronic inflammation, indicated by increased level of pro-inflammatory genes, confers susceptibility to metabolic diseases and it promotes metabolic syndrome. Bile acids were initially considered as lipid solubilizers and they are now recognized as signaling molecules with an anti-inflammatory action (Kazuaki et al. 2013; Keitel et al. 2007).

Recent studies have shown that TGR5 incubate the metabolic function by reducing macrophage activation, thereby blocks inflammation. Immunometabolic disorders such as atherosclerosis and diet-induced obesity can be treated by TGR5 agonists. TGR5 regulates energy expenditure via BA-TGR5-cAMP-D2 cascade in BAT and skeletal muscle (Russell 2003). Indeed, the activation of TGR5 yields a excellent strategy for treatment of type 2 diabetes mellitus and other metabolic disorders (Russell 2003). Moreover, TGR5 causes some side effects in the heart and gallbladder. The supplementation of either bile acids derivatives or synthetic small molecule TGR5 agonists activate TGR5 in the epithelium of the gallbladder leads to the smooth-muscle relaxation, prevents the bile secretion, and remarkably increases gallbladder volume (Li et al. 2011). Hua et al. have observed that ligands of TGR5 modify heart rate and blood pressure in primates (Hua et al. 2016). In the intestinal tract, the localized activation of TGR5 is a promising anti-diabetes mellitus strategy while systemic exposure causes some side effects (Hua et al. 2016). The hypoglycemic effect of TGR5 in the intestine can be achieved by with or without low drug concentrations in BAT or skeletal muscle and its possible side effects in gallbladder and heart could be omitted. A gallbladder filling effect was reduced by low systemic exposure of compound from *Tithonia diversifolia* and it displayed a great hypoglycemic efficacy in normal and diabetic adult male wistar rats (Duan et al. 2015). Chinese Hamster Ovary cell (CHO) and chief cells expressing muscarinic receptors which can be activated by conjugated forms of LCA and DCA and other secondary bile acids, DCA and CDCA are antagonists of the formyl peptide receptor. It is well known that TGR5 is a authentic bile acid receptor due to the fact that it can be activated by multiple form of bile acids (Maruyama et al. 2002). In TGR5, 28% of the ORF region amino acid sequences are identical to human sphingosine-1-phosphate receptor (EDG-1), which is a member of the class a (rhodopsin-like) GPCR family. Most of the endogenous bile acids activate TGR5, with graded potencies. Oleanolic acid (OA), induces TGR5 with close potency to LCA that is an active component from the European olive (*Oleaeuropa*) leaves, but it does not activate the nuclear receptor FXR (Duboc et al. 2014; Lieu et al. 2014). The metabolic and inflammatory disorders are treated by TGR5 ligands, which are derived from multiple steroidal and non-steroidal hormones (Pols 2014).

2.4 Regulation of Glucose Homeostasis

The rise of glucose metabolic products in the circulation always causes an increase of bile acids in the blood. This significant increase of bile acids associated with insulin sensitivity that regulates glucose homeostasis (Thomas et al. 2009). OA from olive leaves selectively activates TGR5 without any effect on FXR (Liu and Wong 2010). In mice models of type 2 diabetes, OA lowers plasma glucose and insulin levels in high-fat diet-induced insulin resistance (Liu and Wong 2010). TGR5 agonists exert their anti-diabetic effects through at least two mechanisms. Firstly, TGR5 increases mitochondrial energy expenditure in BAT and skeletal muscle, and it improves glucose utilization through activation of type 2 iodothyronine deiodinase (D2) and thyroxine (Duboc et al. 2014). Secondly, Intestinal L-cells secretes GLP1 that enhances insulin secretion and suppresses appetite and gastrointestinal transit. GLP-1 secretion is promotes TGR5 agonists (Duboc et al. 2014). These effects reduce increases glucose utilization and decreases blood glucose. GLP-1 secretion is regulated by bile acids and TGR5 selective agonists in the murine enteroendocrine cell line STC-1 (Bala et al. 2014), and TGR5-dependent stimulation of GLP-1 release improves liver and pancreatic function and glucose tolerance in obese mice (Duboc et al. 2014). A mechanistic study found that TGR5 activation increases ATP to ADP ratio, which causes shutting down of ATP-dependent potassium channels (K_{ATP}) and alters calcium influx and these changes consequently leads to the secretion of GLP-1 in STC-1 cells (Duboc et al. 2014). Some of the analogues of bile acids can inhibit reabsorption of bile acids in the enterohepatic circulation that has been associated with type 2 diabetes improvements. (Hansen et al. 2014). In TGR5 knockout mice, the bile acid sequestrants induced GLP-1 release and lower glycaemia (Hansen et al. 2014). Consequently, these glycaemic control property of TGR5 activation provide important message that a possible investigation of alternative drug therapy for type II diabetes.

An enhancement of TGR5 activation leads to cAMP synthesis, which stimulates both the mitogen-activated protein kinase (MAPK) pathway protein kinase A (PKA). PKA phosphorylates cAMP response element binding protein (CREP) and this phosphorylation promotes binding of CREP to its target genes and their expression. As TGR5 plays an important role in energy expenditure and is now an excellent target for metabolic disorders such as obesity, insulin resistant and type 2 mellitus. Moreover, the expression of TGR5 in monocytes can regulate the levels of inflammatory cytokine, which indicates that TGR5 is predominantly involved in the regulation of immune responses in human pathologies (Fiorucci et al. 2009; Fu et al. 2011; Meng et al. 2006). However, TGR5 knockout leads to refractory gallstones in response to the high-fat diet in mice (Li et al. 2011). TGR5 dependent increase of GLP-1 and peptide YY (PYY) secretion leads to suppresses gastric emptying and intestinal transit by promoting nitric oxide release from inhibitory motor neurons, and/or director of smooth muscle of gut (Appleby and Walters 2014). The expression of TGR5 in inhibitory motor neurons of the enteric nervous system is associated with suppression of spontaneous phasic activity of ileum and colon, which was confirmed in isolated segments of ileum and colon treated with DCA (Poole et al. 2010). In gastric smooth muscle, TGR5 receptors are involved in muscle relaxation through the Gαs/cAMP pathway, and it can activate both PKA-dependent phosphorylation of RhoA at Ser^{188} and protein kinase A (PKA)-independent pathways. It also stimulates Ras-related protein 1 (Rap1) signaling by activating exchange protein activated by cAMP.

Rajagopal et al. (2013) have found that TGR5 expression in gastric smooth muscle cells and TGR5 agonists induces smooth muscle relaxation. A potent TGR5 agonist OA, which does not activate the nuclear membrane receptor FXR (Carmen et al. 2017), stimulates muscle relaxation by inducing Gαs activity, increasing production of cAMP and PKA activity (Berdeaux and Stewart 2012). In contrast to endogenous bile acids, a more selective activation of TGR5 by OA indicates its therapeutic significance, which was confirmed by the suppression of OA-induced cAMP formation and muscle relaxation by TGR5 siRNA or $tgr5^{-/-}$ mice.

2.5 TGR5 Activation in Smooth Muscle Physiology

The mechanisms and signaling pathways of TGR5 receptors mediated relaxation of sustained contraction in smooth muscle are depicted by Rajagopal et al., (2013) as shown in Fig. 2.3. TGR5 dependent activation of both PKA and PKG phosphorylates RhoA at serine 188, which leads to RhoA inactivation and its translocation back to the cytosol. This in turn causes reduced activity of RhoA-dependent Rho kinase, which is a positive modulator of ACh-stimulated MLC20 phosphorylation and sustained muscle contraction. It is well documented that Rho Kinase directly inhibits MLC phosphatase to sustain MLC20 phosphorylation (Murthy et al. 2003).

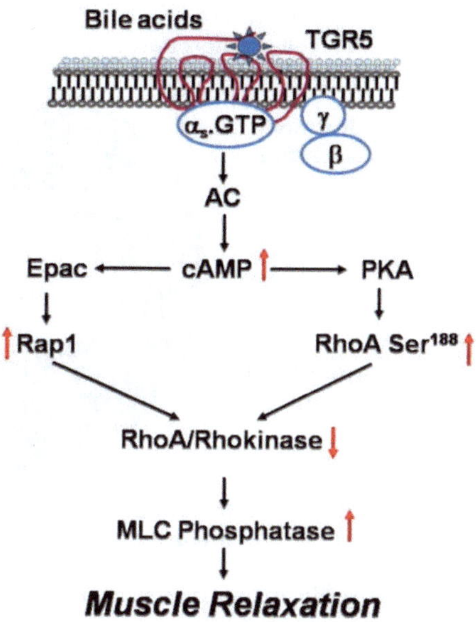

Fig. 2.3 TGR5 activation on smooth muscle physiology

Epacs can also mediate TGR5-Rhokinase signaling in muscle relaxation. Epacs contain a conserved cyclic nucleotide binding domain (CNBD) in N-terminal and a catalytic domain in C-terminal. The catalytic domain is involved in the auto-inhibition of GEF activity, which is relieved by binding of cAMP to the CNBD that induces conformational changes within the protein (Blumenthal et al. 2014).

2.6 Regulations of TGR5 and FXR Receptors

It is well defined that the endocrine functions of bile acids are mediated by nuclear specific bile acid receptor, FXR (NR1H4) and plasma membrane-bound G protein-coupled receptor TGR5 (Bhattacharya et al. 2013; Chang et al. 2016a) (Fig. 2.4).

Among the different types of peripheral tissues, the liver and gut highly express FXR, which is selectively activated by CDCA (Parks et al. 1999; Wang et al. 1999). FXR acts as sensors of metabolic signals that controls bile acid, glucose and cholesterol homeostasis by modulating the expression of genes associated with specific metabolic pathways (Chaube et al. 2014; Jiao et al. 2015). FXR could regulates bile acids biosynthesis through negative-feedback control on Cholesterol 7-α hydroxlase (CYP7A1), which is a main enzyme involved in the initiation of bile acids biosynthesis (Russell 2009). When the blood glucose levels are increased, FXR blocks gluconeogenesis and promotes glycogen biosynthesis in liver. LCA is a more selective ligand of TGR5 (Keitel et al. 2010; Wang et al. 2011), which TGR5 controls the insulin release in pancreas via direct activation of β cells in the pancreas and its indirect effect on the release of GLP-1 and glucose-dependent insulinotropic peptide (GIP) from enteroendocrine cells (Chang et al. 2016a; Song et al. 2000) (Fig. 2.5).

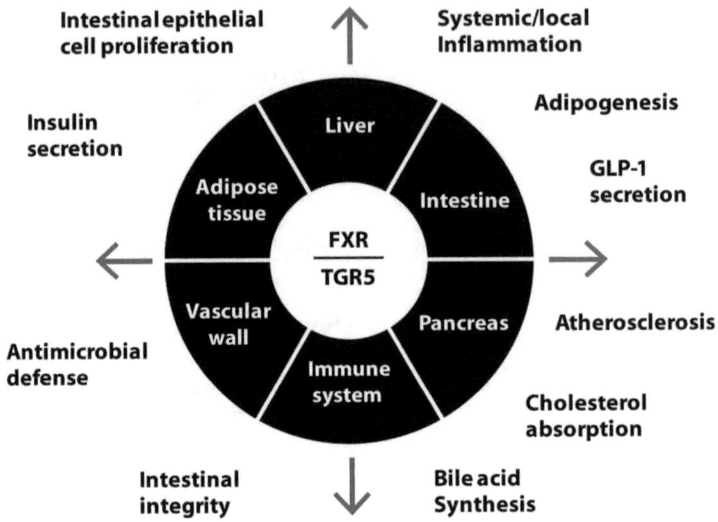

Fig. 2.4 Regulations of plasma membrane receptor (TGR5) and nuclear receptor (FXR)

Receptor	FXR	TGR5
Ligands*	CDCA > LCA, DCA > CA	LCA > DCA > CDCA > CA
Liver cell expression	Hepatocyte, stellate cell, Sinusoid endothelial cell	Sinusoid endothelial cell Kupffer cell, hepatocyte
Functions in liver	BA metabolism, antiapoptosis Anti-inflammatory response, liver regeneration, HCC suppression	Suppress LPS response, glucagon expression, energy metabolism, liver inflammation, and HCC?

* CDCA: chenodeoxycholic acid; CA: cholic acid; DCA: deoxycholic acid; LCA: lithocholic acid; LPS: lipopolysaccharide.

Fig. 2.5 Ligands of plasma membrane receptor (TGR5) and nuclear receptor (FXR)

2.7 Functions of Gut Microbiota in Bile Acid Homeostasis

The gut-to-liver axis is the important steps in bile acid metabolism and gut microbiota controls enterohepatic circulation of bile acids, synthesis and bile acid pool size and composition; whereas, the gut microbiota's over growth is controlled by bile acids. Reshaping the gut microbiome is very important steps in host metabolism and metabolic diseases and this reshape is regulates by Bile acids (Zheng et al. 2017). There are controversial reports about the regulation of FXR, which inhibits synthesis of bile acid and fatty acid and it enhances glucose and insulin sensitivity in obese and diabetic mice (Cipriani et al. 2010). Commonly, the activation of FXR reduces bile acid pool and progressive fibrosis in liver, which is good for primary biliary cirrhosis and non-alcoholic steatohepatitis (NASH) (Arab et al. 2017).

The intestinal activation of FXR and TGR5 enhances insulin sensitivity and it stimulates TGR5-mediated GLP-1 secretion (Chiang 2017a). FXR agonist GW4064 decrease proglucagon expression in intestinal L cells by interfering with glucose-responsive carbohydrate-response element binding protein and GLP-1 secretion (Mohamed-Sami et al. 2017). It is well known that bile acids are involved in the stimulation of glucose-induced GLP-1 secretion. However, the paradoxical effects of FXR agonists on glucose tolerance and insulin sensitivity could possibly due to variation in standard condition of experimental animal maintenance and genetic background as well as differential effects of FXR agonists on gut microbiota.

2.8 Bile Acid Signalling in Metabolic Regulation

A known endogenous agonist, bile acids activates both the nuclear receptor FXR, membrane TGR5 and Sphingosine-1-Phosphate Receptor 2 (S1PR2) in the gastrointestinal systems, which are studied almost last two decades. Despite the roles of FXR and TGR5 in the regulation of bile acid synthesis and homeostasis are well defined in human and experimental animal models, the physiological role of bile acid signaling pathways under normal conditions are not completely understood. In fact, a majority of studies have been performed in mice, which have a different bile acid composition and pool size compared with humans (Proia and Hla 2015) (Fig. 2.6).

TGR5 enhances the GLP-1 release in the enteroendocrine cells and it is expressed by mouse and human pancreatic β cell lines in which it has insulinotropic effects (Taoufiq et al. 2012). OA and INT-777 are more selective ligands to TGR5 that are

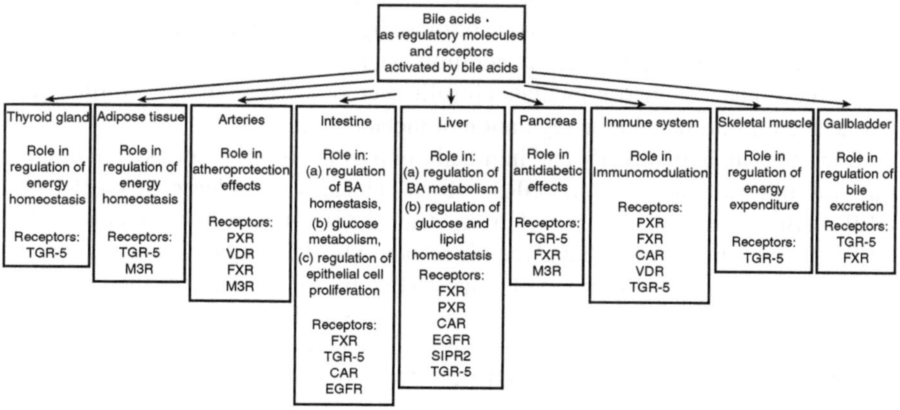

Fig. 2.6 Regulations of Bile acids as regulatory molecules

selectively activate Gαs and downstream signalling by increasing intracellular cAMP and Ca^{2+}. The primary function of bile salts, the amphipathic molecules, are intestinal micelle formation and lipid digestion (Hofmann and Eckmann 2006). In recent years, they have extended far beyond lipid digestion that includes the modulation of cholesterol and triglyceride metabolism, insulin sensitivity, the intestinal endocrine response to meals, and energy homeostasis (Inagaki et al. 2006; Kumar et al. 2012), and many of these functions are mediated by the FXR and TGR5. FXR is abundantly expressed in liver, adipose tissue and muscle and it is the key negative regulator of bile acid synthesis (Ding et al. 2015; Pellicciari et al. 2005).

TGR5s controls the different physiological functions in body such as hepatic blood flow, inhibits the inflammation, mobilization of intestinal and gallbladder, and the activity of skeletal muscle iodothyronine deiodinase (D2), which is a key controller of metabolic homeostasis (Drigo et al. 2013; Heemstra et al. 2009). GLP-1 and GLP-2 from the pancreas β-cells are vital factors of the intestinal endocrine response to meals and that stimulate insulin release (Bell et al. 1983; Yutaka et al. 2010). Glucose concentration in the circulation is crucial factor for insulin secretion, which is crucial under basal and post-prandial conditions for the energy homeostasis. The glucose metabolism within the β cell increases the ATP: ADP ratio, which is an initial step for the insulin release. The increased ATP-ADP ratio inhibits ATP sensitive K^+ channels, which consequently leads to plasma membrane depolarization and opening of voltage-gated Ca^{2+} channels (Ashcroft 2005; MacDonald et al. 2002). The influx of Ca^{2+} and cytosolic increase of Ca^{2+} initiates exocytosis process and release of insulin. In general, glucose-dependent insulin secretion pathway is modulated by several paracrine and endocrine mechanisms.

Several researchers found that the expression of FXR and TGR in β cells associated with insulin release (Dufer et al. 2012), which is one of the centre-pieces of the body's strategy to maintain metabolic homeostasis. Thus, it is not surprising that pancreatic islet β cell function is modulated by cell–cell interactions as well as neural, endocrine and paracrine factors (Franklin and Wollheim 2004; Patrik et al. 2012). Wide varieties of endogenous molecules bind their specific receptors on islet cells and activate/integrate various signals to modulate the synthesis and secretion. Bile salts are recently identified as one of those ligands that influence islet function, in particular, via binding to FXR. In addition, islet β cells also express TGR5 receptors, which enhance glucose-mediated insulin secretion. These findings further support a role for bile salts in the regulation of metabolism.

Bile salts can both directly and indirectly impact the islet functions. In the intestine, the bile salts activate of TGR5, which enhances the release of GLP-1 and GLP-2 (Brighton et al. 2015; Lefebvre et al. 2009). It is well documented that GLP-1 flattens the postprandial glycemic response by delaying the gastric emptying and thus it reduces the absorption of glucose (Lefebvre et al. 2009). GLP-1 also acts as a tropic factor and promotes insulin secretion from islet cells following a glucose challenge (Hisae et al. 2017). As like many other insulin secretion stimulators, TGR5 ligands such as LCA, OA and INT 777 activate final exocytosis of insulin

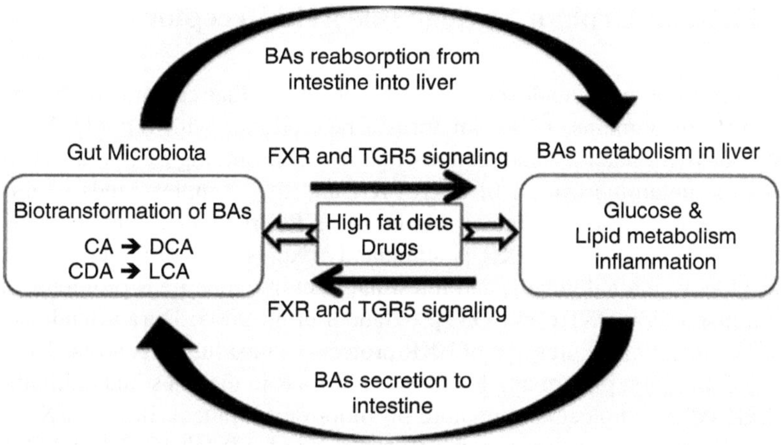

Fig. 2.7 FXR and TGR5 signalling in physiological functions

through classical Ca^{2+}-dependent pathways, which is evident from increased level of intracellular Ca^{2+} upon TGR5 activation and insulin secretion blunted by calcium chelation. The level of blood glucose is a key factor that drives the insulin secretion. The islet cells sense ambient glucose, fatty acids and various amino acids levels, which activate pro-insulin synthesis and insulin secretory mechanisms in response to the metabolic coupling factors (Kumar et al. 2012; Nicholls 2016). The levels of bile salts in the circulation are generally less than 5 μM and it increases up to 15 μM in post-prandial conditions (Monte et al. 2009; Trottier et al. 2011). The specific effects of bile salts *in vivo* are mainly depending on its biochemical nature in addition to their concentration. For example, tauro-ursodeoxycholic acid is a weak ligand for TGR5 and FXR, whereas LCA and CDCA are more potent ligands. In chronic pathological conditions in the liver such as cholestasis, the bile salts levels are highly increased in fasting condition and there is a loss of secondary bile salts (Arab et al. 2017; Ulrich et al. 2015).

A decrease in the intestinal exposure to bile salts could also dampen the incretin response to meals, which is a vital hormone required for islet function and glucose homeostasis. In public health point of view, this dysfunction could contribute to the development of type 2 diabetes, a leading threat to the human. The insulin resistance and type 2 diabetes is characterised by hyperinsulinemia under fasting condition and eventual β cell exhaustion (Cerf 2013). In this case, a combination of persistent exposure of β cells to inflammatory cytokines, high levels of fatty acids and other lipolysis products accompanied by increased blood sugar drive excessive insulin synthesis and secretion (Röder et al. 2016). The link between insulin secretory response to these factors and bile salts in the genesis of type 2 diabetes as well as its role as a disease modifier awaits a detailed elucidation (Fig. 2.7).

2.9 FXR, an Orphan Nuclear Bile Acid Receptor

FXR is a multipurpose nuclear receptor super family that contains both DNA and ligand binding domains. FXR can form a heterodimer with retinoid X receptor (RXR) that is initiated by endogenous ligands. The binding of bile acids to FXR initiates the heteromerization of RXR/FXR and this complex binds to the FXR response elements in the promoter regions of FXR target genes and enhances their expression (Chiang 2002). FXR is abundantly expressed by the hepatic tissue as well as kidney, adrenal glands, and intestinal tract in which FXR promotes organic solute transporters, OSTα and OSTβ (Claudel et al. 2005; Duran-Sandoval et al. 2005). The intestinal expression of FXR protects against tumorigenesis. The modulation of expression pattern and localisation of FXR in the intestinal epithelial cells is considered as a molecular signature of tumor development. In fact, FXR acts as an intracellular bile acid sensor and it is involved in feedback regulation of bile acid synthesis (Hofmann 1999; Modica et al. 2008). Emerging evidences indicate that FXR play an important role in various biological functions including the function of pancreatic β cells, regulation of hepatic glucose production, liver regeneration, lipid metabolism, intestinal homeostasis, and urine concentration in the kidney. Peroxisome proliferator-activated receptor-gamma coactivator 1 alpha (PGC-1 alpha) induced expression and activation of FXR is required for the regulation of triglyceride metabolism (Nierenberg et al. 2018) (Fig. 2.8).

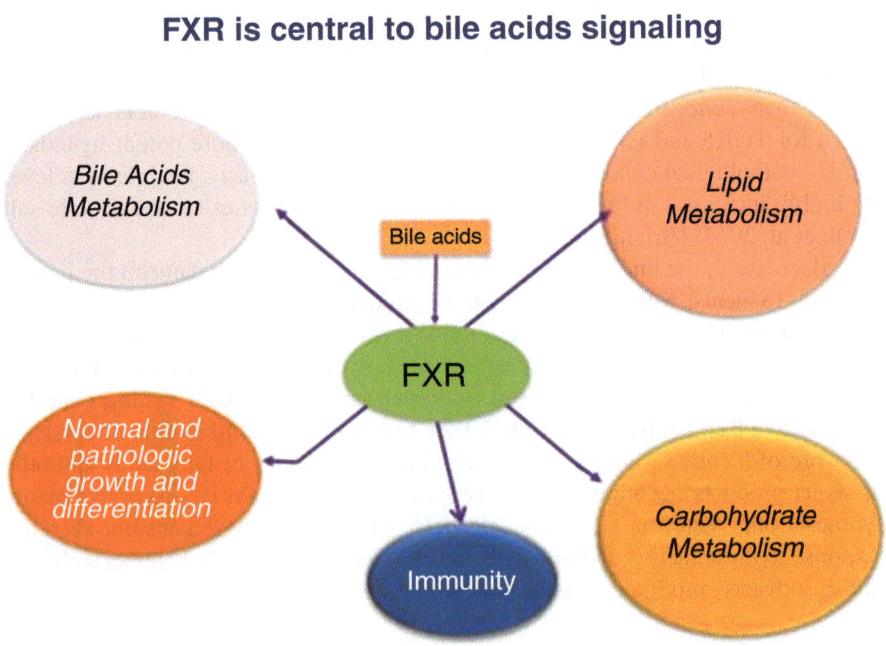

Fig. 2.8 FXR signaling in physiological functions

2.9.1 FXR Regulates Glucose/Lipid Metabolism

FXR has an essential role in controlling carbohydrate metabolism by regulating the expression of phosphoenolpyruvate carboxykinase (PEPCK), which is a key enzyme involved in hepatic gluconeogenesis (Chiang 2002; Jacinto and Fang 2014). The impairment of glucose tolerance and insulin resistance in FXR-null mice clearly shows the crucial role of hepatic FXR in glucose homeostasis. In addition to governing the expression/activity of PEPCK and glucose-6-phosphatase (G6pc), FXR also positively modulates the glycogen synthesis and increases its level in hepatocytes via enhancing downstream insulin receptor signalling (Chang et al. 2016b). An extensive studies in FXR-null mice showed that FXR also affects lipid metabolism.

In FXR-null mice, the levels of plasma triglycerides and cholesterol are remarkably increased. In addition, the genetic ablation of FXR in mice exhibits increased level high density lipoprotein (HDL) cholesterol in the circulation, which could be due to reduction in the hepatic expression of scavenger receptor class B member 1 (SCARB1), a receptor facilitating the clearance of HDL cholesterol from the circulation. However, the activation FXR using its agonist exhibited less plasma triglycerides in wild type mice compared to FXR-null mice. In addition, the activation of FXR or over expression of FXR improves hyperglycemia and hyperlipidemia by lowering blood glucose in experimental animal model of diabetes. Furthermore, FXR can also directly influence the lipoprotein metabolism by regulating a set of genes including SREBP-1c, phospholipid transfer protein, SCD-1, the very low density lipoprotein receptor, apolipoprotein C-II, and apolipoprotein E (Chang et al. 2016b; Chiang 2017b; Sirvent et al. 2004). These studies reveal that FXR is a critical regulator of lipid metabolism by governing the cholesterol homeostasis as well as synthesis of fatty acids and triglycerides.

2.9.2 FXR Signalling in Liver Regeneration

The liver regeneration following an injury and various insults are triggered by sequential changes in gene expressions induced by specific stimuli and morphologic reconstruction. At cellular level, these adaptive responses for the functional recovery are orchestrated by a highly complex network of signal transductions in hepatocytes. Bile acids act as key metabolic signalling modulators during liver regeneration following a 70% partial hepatectomy by activating FXR downstream pathways, which indicates that the bile acid signalling mediated by FXR favours the reconstruction of the liver (Jacinto and Fang 2014; Wang et al. 1999; Zhang et al. 2003).

2.9.3 FXR Regulation of Pancreatic β Cells Physiology

The expression and activation of FXR in pancreatic beta cells indispensable for the systemic glucose homeostasis. FXR activated by bile acids stimulates insulin secretion in beta-cells by inhibiting K (ATP) channel. FXR is localized in the cytosol in lean mice, where as FXR is predominately in the nucleus of obese mice and this receptor is translocated into nucleus upon activation of ligand. However, the activation of FXR triggers the production and secretion of insulin by increasing Ca^{2+}-influx and reducing the KATP current. FXR also protects human islets from lipotoxicity. As FXR controls K_{ATP} currents in pancreatic β cells, its act as a cytosolic and non-genomic effectors in those cells., All these findings reveal that the activation of FXR most effectively restores restore glucose-dependent insulin secretion in pancreatic β cells and it could be one of the promising strategies for treating T2D (Ding et al. 2015; Dufer et al. 2012; Düfer et al. 2012; Iuliana Ristea et al. 2010).

2.10 Summary

Bile acid induced TGR5/FXR receptors regulate liver and intestinal metabolism, gut microbiome, host metabolism, and diseases. Increasing evidences indicate that components of bile acid homeostasis have been translated to drug therapy for various physiological and pathological disorders. The beneficiary and most effective therapeutic responses of FXR and TGR5 receptors can be achieved by the development of novel synthetic agonists for those receptors. As these receptors could be utilised as a therapeutic targets for a variety if metabolic disorders including carbohydrate and lipid metabolic disorders, a clear and deep understanding of the roles of FXR/TGR5 in all systems will hasten the development of most efficient FXR/TGR5 ligands/modulators for the treatment of metabolic disorders.

References

Appleby, R., & Walters, J. (2014). The role of bile acids in functional GI disorders. *Neurogastroenterology and Motility, 26*, 1057–1069.

Arab, J., Karpen, S., Dawson, P., Arrese, M., & Trauner, M. (2017). Bile acids and nonalcoholic fatty liver disease: Molecular insights and therapeutic perspectives. *Hepatology, 65*, 350–362.

Ashcroft, F. (2005). ATP-sensitive potassium channelopathies: Focus on insulin secretion. *Journal of Clinical Investigation, 115*, 2047–2058.

Bala, V., Mahavadi, S., Rajagopal, S., Zhou, R., Kuemmerle, J. F., Sanyal, A. J., & Murthy, K. S. (2011). Bile acid-induced stimulation of ERK1/2 activity, GLP-1 and PYY release in enteroendocrine cells are mediated by the activation of Epac/PLC-ε signaling pathway via GS-coupled TGR5. *Gastroenterology, 140,* S–147.

Bala, V., Rajagopal, S., Kumar, D. P., Nalli, A. D., Mahavadi, S., Sanyal, A. J., Grider, J. R., & Murthy, K. S. (2014). Release of GLP-1 and PYY in response to the activation of G-protein coupled bile acid receptor TGR5 is mediated by Epac/PLC e pathway and modulated by endogenous H_2S. *Frontiers in Physiology, 5*, 1–11.

Bell, G., Santerre, R., & Mullenbach, G. (1983). Hamster preproglucagon contains the sequence of glucagon and two related peptides. *Nature, 302*, 716–718.

Berdeaux, R., & Stewart, R. (2012). cAMP signaling in skeletal muscle adaptation: Hypertrophy, metabolism, and regeneration. *Amercan Journal of Physiology; Endocrinology and Metabolism, 303*, E1–E17.

Bhattacharya, S., Mahavadi, S., Al-Shboul, O., Rajagopal, S., Grider, J. R., & Murthy, K. S. (2013). Differential regulation of muscarinic m2 and m3 receptor signaling in gastrointestinal smooth muscle by caveolin-1. *American Journal of Physiology; Cell Physiology, 305*, C334–C347.

Blumenthal, D., Copps, J., Smith-Nguyen, E., Zhang, P., Heller, W., & Taylor, S. (2014). The roles of the RIIβ linker and N-terminal cyclic nucleotide-binding domain in determining the unique structures of the Type IIβ. *Journal of Biological Chemistry, 289*, 28505–28512.

Brighton, C., Rievaj, J., Kuhre, R., Glass, L., Schoonjans, K., Holst, J., Gribble, F., & Reimann, F. (2015). Bile acids trigger GLP-1 release predominantly by accessing basolaterally located G Protein–coupled bile acid receptors. *Endocrinology, 156*, 3961–3970.

Broeders, E., Nascimento, E., Havekes, B., Brans, B., Roumans, K., Tailleux, A., Schaart, G., Kouach, M., Charton, J., Deprez, B., Bouvy, N., Mottaghy, F., Staels, B., van Marken Lichtenbelt, W., & Schrauwen, P. (2015). The bile acid chenodeoxycholic acid increases human brown adipose tissue activity. *Cellular Metabolsim, 22*, 418–426.

Carmen, F., De Marino, S., Carino, A., Sepe, V., Marchianò, S., Cipriani, S., Di Leva, F., Limongelli, V., Monti, M., Capolupo, A., Distrutti, E., Fiorucci, S., & Zampella, A. (2017). Targeting bile acid receptors: Discovery of a potent and selective farnesoid X receptor agonist as a new lead in the pharmacological approach to liver diseases. *Frontiers in Pharmacology, 8*, 162.

Cerf, M. (2013). Beta cell dysfunction and insulin resistance. *Frontiers in Endocrinology, 4*, 37.

Chang, H., Chou, C., Lin, Y., Shieh, P., Kuo, D., Jan, C., & Liang, W. (2016a). Esculetin, a natural coumarin compound, evokes Ca(2+) movement and activation of Ca(2+)-associated mitochondrial apoptotic pathways that involved cell cycle arrest in ZR-75-1 human breast cancer cells. *Tumour Biology, 37*, 4665–4678.

Chang, Y., Tae, H., Kim Ja, H., & Geon, K. (2016b). Farnesoid X receptor as a regulator of fuel consumption and mitochondrial function. *Archieves of Pharmacological Research, 39*, 1062–1074.

Chaube, R., Hess, D., Wang, Y., Plummer, B., Sun, Q., Laurita, K., & Stamler, J. (2014). Regulation of the skeletal Muscle ryanodine receptor/Ca^{2+}-release channel RyR1 by S-palmitoylation. *Journal of Biological chemistry, 289*, 8612–8619.

Chen, X., Lou, G., Meng, Z., & Huang, W. (2011). TGR5: A novel target for weight maintenance and glucose metabolism. *Experimental Diabetes and Research, 2011*, 1–5.

Chiang, J. (2002). Bile acid regulation of gene expression: Roles of nuclear hormone receptors. *Endocrinology Review, 23*, 443–463.

Chiang, J. (2017a). Bile acid metabolism and signaling in liver disease and therapy. *Liver Research, 1*, 3–9.

Chiang, J. (2017b). Recent advances in understanding bile acid homeostasis. *F1000 Research, 6*, 2029.

Cipriani, S., Mencarelli, A., Palladino, G., & Fiorucci, S. (2010). FXR activation reverses insulin resistance and lipid abnormalities and protects against liver steatosis in Zucker (fa/fa) obese rats. *Journal of Lipid Research, 51*, 771–784.

Claudel, T., Staels, B., & Kuipers, F. (2005). The farnesoid X receptor: A molecular link between bile acid and lipid and glucose metabolism. *Arteriosclerosis Thrombosis and Vascular Biology, 25*, 2020–2030.

Ding, L., Yang, L., Zhengtao, W., & Huangb, W. (2015). Bile acid nuclear receptor FXR and digestive system diseases. *Acta Pharmacologica Sinica, 5*, 135–144.

Drigo, R., Fonseca, T., Werneck-de-Castro, J., & Bianco, A. (2013). Role of the type 2 iodothyronine deiodinase (D2) in the control of thyroid hormone signaling. *Biochimica Biophysica Acta, 1830*, 3956–3964.

Duan, H., Ning, M., Zou, Q., Ye, Y., Feng, Y., Zhang, L., Leng, Y., & Shen, J. (2015). Discovery of intestinal targeted TGR5 agonists for the treatment of type 2 diabetes. *Journal of Medicinal Chemistry, 58*, 3315–3328.

Duboc, H., Taché, Y., & Hofmann, A. (2014). The bile acid TGR5 membrane receptor: From basic research to clinical application. *Digestive Liver Diseases, 46*, 302–312.

Dufer, M., Hörth, K., Krippeit-Drews, P., & Drews, G. (2012). The significance of the nuclear farnesoid X receptor (FXR) in beta cell function. *Islets, 4*, 333–338.

Düfer, M., Hörth, K., Wagner, R., Schittenhelm, B., Prowald, S., Wagner, T., Oberwinkler, J., Lukowski, R., Gonzalez, F., Krippeit-Drews, P., & Drews, G. (2012). Bile acids acutely stimulate insulin secretion of mouse β-cells via farnesoid X receptor activation and K(ATP) channel inhibition. *Diabetes, 61*, 1479–1489.

Duran-Sandoval, D., Cariou, B., Percevault, F., Hennuyer, N., Grefhorst, A., van Dijk, T., Gonzalez, F., Fruchart, J., Kuipers, F., & Staels, B. (2005). The farnesoid X receptor modulates hepatic carbohydrate metabolism during the fasting-refeeding transition. *Journal of Biological Chemistry, 280*, 29971–29979.

Fiorucci, S., Mencarelli, A., Palladino, G., & Cipriani, S. (2009). Bile-acid-activated receptors: Targeting TGR5 and farnesoid-X-receptor in lipid and glucose disorders. *Trends in Pharmacological Sciences, 30*, 570–580.

Foord, S., Bonner, T., Neubig, R., Rosser, E., Pin, J., Davenport, A., Spedding, M., & Harmar, A. (2005). International union of pharmacology. XLVI. G protien coupled receptor list. *Pharmacological Reviews, 57*, 279–288.

Franklin, I., & Wollheim, C. (2004). GABA in the endocrine pancreas. *Journal of General Physiology, 123*, 185–190.

Fu, D., Wakabayashi, Y., Lippincott-Schwartz, J., & Arias, I. (2011). Bile acid stimulates hepatocyte polarization through a cAMP-Epac-MEKLKB1-AMPK pathway. *Proceeding of Natural Acadmics of Sciences, USA, 108*, 1403–1408.

Guo, C., Su, J., Li, Z., Xiao, R., Wen, J., Li, Y., Zhang, M., Zhang, X., Yu, D., Huang, W., Chen, W., & Wang, Y. (2015). The G-protein-coupled bile acid receptor Gpbar1 (TGR5) suppresses gastric cancer cell proliferation and migration through antagonizing STAT3 signaling pathway. *Oncotarget, 6*, 34402–34413.

Hansen, M., Sonne, D., & Knop, F. (2014). Bile acid sequestrants: Glucose-lowering mechanisms and efficacy in type 2 diabetes. *Current Diabetes Reports, 14*, 485–495.

Heemstra, K., Soeters, M., Fliers, E., Serlie, M., Burggraaf, J., van Doorn, M., van der Klaauw, A., Romijn, J., Smit, J., Corssmit, E., & Visser, T. (2009). Type 2 iodothyronine deiodinase in skeletal muscle: Effects of hypothyroidism and fasting. *Journal of Clinical Endocrinology and Metabolism, 94*, 2144–2150.

Hiroki, T., Yoko, Y., Kohkichi, M., Naho, K., Tatsuya, T., Yoko, T., Kazuo, T., & Mitsuhiro, W. (2016). Role of bile acids in the regulation of the metabolic pathways. *World Journal of Diebetes, 7*, 260–270.

Hisae, A., Koro, G., Fujiwara, K., Anai, M., Chiba, S., Masaki, T., Kakuma, T., & Shibata, H. (2017). Glucagon-like peptide-1 reduces pancreatic β-cell mass through hypothalamic neural pathways in high-fat diet-induced obese rats. *Scientific Reports, 7*, 5578.

Hofmann, A. (1999). The continuing importance of bile acids in liver and intestinal disease. *Archieves of Internal Medicine, 159*, 2647–2658.

Hofmann, A., & Eckmann, L. (2006). How bile acids confer gut mucosal protection against bacteria. *Proceeding of Natural Acadmics of Sciences, USA, 103*, 4333–4334.

Houten, S., Watanabe, M., & Auwerx, J. (2006). Endocrine functions of bile acids. *EMBO Journal, 25*, 1419–1425.

Hua, C., Zhi-Xiang, C., Wang, K., Meng-Meng, N., Qing-An, Z., Ying, F., Yang-Liang, Y., Ying, L., & Jian-Hua, S. (2016). ntestinally-targeted TGR5 agonists equipped with quaternary ammonium have an improved hypoglycemic effect and reduced gallbladder filling effect. *Scientific Reports, 6*, 1–6.

Inagaki, T., Moschetta, A., Youn-Kyoung, L., Peng, L., Zhao, G., Downes, M., Yu, R., Shelton, J., Richardson, J., Repa, J., Mangelsdorf, D., & Kliewer, S. (2006). Regulation of antibacterial defense in the small intestine by the nuclear bile acid receptor. *Proceeding of Natural Acadmics of Sciences, USA, 103*, 3920–3925.

Iuliana Ristea, P., Helleboid-Chapman, A., Lucas, A., Vandewalle, B., Dumont, J., Bouchaert, E., Derudas, B., Kerr-Conte, J., Caron, S., Pattou, F., & Staels, B. (2010). The nuclear receptor FXR is expressed in pancreatic β-cells and protects human islets from lipotoxicity. *FEBS Letter, 584*, 2845–2851.

Jacinto, S., & Fang, S. (2014). Essential roles of bile acid receptors FXR and TGR5 as metabolic regulators. *Animal cells and Systems, 18*, 359–364.

Jiao, Y., Lu, Y., & Li, X. (2015). Farnesoid X receptor: A master regulator of hepatic triglyceride and glucose homeostasis. *Acta Pharmacologica Sinica, 36*, 44–50.

Katsuma, S., Hirasawa, A., & Tsujimoto, G. (2005). Bile acids promote glucagon-like peptide-1 secretion through TGR5 in a murine enteroendocrine cell line STC-1. *Biochemical and Biophysical Research Communication, 329*, 386–390.

Kawamata, Y., Fujii, R., Hosoya, M., Harada, M., Yoshida, H., Miwa, M., Fukusumi, S., Habata, Y., Itoh, T., Shintani, Y., Hinuma, S., Fujisawa, Y., & Fujino, M. (2003). A G protein-coupled receptor responsive to bile acids. *Journal of Biological Chemistry, 278*, 9435–9440.

Kazuaki, Y., Tadakazu, H., Katsuyoshi, S., Nobuhiko, K., Riko, I., Kitazume, M., Maiko, M., Michihide, U., Yuka, N., Katsuyoshi, M., Toshiro, S., Kazutaka, K., Akira, S., Takanori, K., & Toshifumi, H. (2013). TGR5 signalling inhibits the production of pro-inflammatory cytokines by in vitro differentiated inflammatory and intestinal macrophages in Crohn's disease. *Immunology, 139*, 19–29.

Keitel, V., Reinehr, R., Gatsios, P., Rupprecht, C., Görg, B., Selbach, O., Häussinger, D., & Kubitz, R. H. (2007). The G-protein coupled bile salt receptor TGR5 is expressed in liver sinusoidal endothelial cells. *Hepatology, 45*, 695–704.

Keitel, V., Gorg, B., Bidmon, H., Zemtsova, I., Spomer, L., Zilles, K., & Haussinger, D. (2010). The bile acid receptor tgr5 (gpbar-1) acts as a neurosteroid receptor in brain. *Glia, 58*, 1794–1805.

Kida, T., Tsubosaka, Y., Hori, M., Ozaki, H., & Murata, T. (2013). Bile acid receptor TGR5 agonism induces NO production and reduces monocyte adhesion in vascular endotheilal cells. *Arteriosclerosis Thrombosis and Vascular Biology, 33*, 1663–1669.

Kumar, D., Senthilkumar, R., Sunila, M., Faridoddin, M., Grider, J., Murthy, K., & Sanyal, A. (2012). Activation of transmembrane bile acid receptor TGR5 stimulates both insulin gene transcription and insulin release in pancreatic b cells. *Biochemical and Biophysical Research Communication, 427*, 600–605.

Lefebvre, P., Cariou, B., Lien, F., Kuipers, F., & Staels, B. (2009). Role of bile acids and bile acid receptors in metabolic regulation. *Physiology Review, 89*, 147–191.

Li, T., SR, H., Kir, S., Umetani, M., Schmidt, D., Kliewer, S., & Mangelsdorf, D. (2011). The G Protein-coupled bile acid receptor, TGR5, stimulates gallbladder filling. *Molecular Endocrinology, 25*, 1066–1071.

Lieu, T., Jayaweera, G., Zhao, P., Poole, D., Jensen, D., Grace, M., McIntyre, P., Bron, R., Wilson, Y., Krappitz, M., Haerteis, S., Korbmacher, C., Steinhoff, M., Nassini, R., Materazzi, S., Geppetti, P., Corvera, C., & Bunnett, N. (2014). The bile acid receptor TGR5 activates the TRPA1 channel to induce itch in mice. *Gastroenterology, 147*, 1417–1428.

Liu, W., & Wong, C. (2010). Oleanolic acid is a selective farnesoid X receptor modulator. *Phytotherapy Research, 24*, 369–373.

MacDonald, P., Sewing, S., Wang, J., Joseph, J., Smukler, S., Wang, J., Monique, C., Catherine, B., Tsushima, R., Salapatek, A., & Wheeler, M. (2002). Inhibition of Kv2.1 voltage-dependent K+channels in pancreatic β-cells enhances glucose-dependent insulin secretion. *Journal of Biological Chemistry, 277*, 44938–44945.

Maruyama, T., Miyamoto, Y., Nakamura, T., Tamai, Y., Okada, H., Sugiyama, E., Nakamura, T., Itadani, H., & Tanaka, K. (2002). Identification of membrane-type receptor for bile acids (M-BAR). *Biochemical and Biophysical Research Communication, 298*, 714–719.

Masyuk, A., Huang, B., Radtke, B., Gajdos, G., Splinter, P., Masyuk, T., Gradilone, S., & LaRusso, N. (2013). Ciliary subcellular localization of TGR5 determines the cholangiocyte functional response to bile acid signaling. *American Journal of Physiology: Gastrointestinal and Liver Physiology, 304*, G1013–G1024.

McMillin, M., Gabriel, F., Richard, T., Giuseppina, D., Jenny, S., Hope, S., Newell-Rogers, K., Grant, S., & DeMorrow, S. (2015). TGR5 signaling reduces neuroinflammation during hepatic encephalopathy. *Journal of Neurochemistry, 135*, 565–576.

Meng, J., Ceryak, S., Aratsu, Z., Jones, L., Epstein, L., & Bouscarel, B. (2006). Biphasic regulation by bile acids of dermal fibroblast proliferation through regulation of cAMP production and COX-2 expression level. *American Journal of Physiology; Cell Physiology, 291*, C546–C549.

Modica, S., Murzilli, S., Salvatore, L., Schmidt, D., & Moschetta, A. (2008). Nuclear bile acid receptor FXR protects against intestinal tumorigenesis. *Cancer Research, 68*, 9589–9594.

Mohamed-Sami, T., Sophie, L., Bart, S., & Xavier, C. (2017). Intestinal bile acid receptors are key regulators of glucose homeostasis. *Procedings of the Nutrition Society, 76*, 192–202.

Monte, M., Marin, J., Antelo, A., & Vazquez-Tato, J. (2009). Bile acids: Chemistry, physiology, and pathophysiology. *World Journal of Gastroenterology, 15*, 804–816.

Murthy, K., Zhou, H., Grider, J., & Makhlouf, G. (2003). Inhibition of sustained smooth muscle contraction by PKA and PKG preferentially mediated by phosphorylation of RhoA. *Amercan Journal of Physiology; Gastrointestinal Liver Physiology, 284*, G1006–G1016.

Nicholls, D. (2016). The pancreatic -cell: A bioenergetic perspective. *Physiology Review, 96*, 1385–1447.

Nierenberg, A., Ghazna, S., Sadora Sande, I., Ellarda, K., Janosa, J., & Sylvia, L. (2018). Peroxisome proliferator-activated receptor gamma coactivator-1 alpha as a novel target for bipolar disorder and other neuropsychiatric disorders. *Biological Psychiatry, 83*(9), 761–769.

Parks, D., Blanchard, S., Bledsoe, R., Chandra, G., Consler, T., Kliewer, S., Stimmel, J., Willson, T., AM, Z., Morre, D., & Lehmann, J. (1999). Bile acids: Natural ligands for an orphan nuclear receptor. *Science, 284*, 1365–1368.

Patrik, R., Matthias, B., & Zhang, Q. (2012). Regulation of calcium in pancreatic α- and β-cells in health and disease. *Cell Calcium, 51*, 300–308.

Pellicciari, R., Costantino, G., & Fiorucci, S. (2005). Farnesoid X receptor: From structure to potential clinical applications. *Journal of Medcinal Chemistry, 48*, 5383–5403.

Per-Arne, S., Maja, O., CAA, J., Magdalena, T., Pereira, M., Froguel, P., & Jacobson, P. (2013). The TGR5 gene is expressed in human subcutaneous adipose tissue and is associated with obesity, weight loss and resting metabolic rate. *Biochemical and Biophysical Research Communication, 433*, 563–566.

Perino, A., Pols, T., Nomura, M., Stein, S., Pellicciari, R., & Schoonjans, K. (2014). TGR5 reduces macrophage migration through mTOR induced C/EBPbeta differential translation. *Journal of Clinical Investigation, 124*, 5424–5436.

Pols, T. (2014). TGR5 in inflammation and cardiovascular disease. *Biochemical Society Transactions, 42*, 244–249.

Pols, T., Nomura, M., Harach, T., Lo Sasso, G., Oosterveer, M., Thomas, C., Rizzo, G., Gioiello, A., Adorini, L., Pellicciari, R., Auwerx, J., & Schoonjans, K. (2011). TGR5 activation inhibits atherosclerosis by reducing macrophage inflammation and lipid loading. *Cellular Metabolsim, 14*, 747–757.

Poole, D., Godfrey, C., Cattaruzza, F., Cottrell, G., Kirkland, J., Pelayo, J., Bunnett, N., & Corvera, C. (2010). Expression and function of the bile acid receptor GpBAR1 (TGR5) in the murine enteric nervous system. *Neurogastroenterology and Motility, 22*, 814–825.

Proia, R., & Hla, T. (2015). Emerging biology of sphingosine-1-phosphate: Its role in pathogenesis and therapy. *Journal of Clinical Investigation, 125*, 1379–1387.

Rajagopal, S., Kumar, D., Mahavadi, S., Bhattacharya, S., Zhou, R., Corvera, C., Bunnett, N., JR, G., & KS, M. (2013). Activation of G protein-coupled bile acid receptor, TGR5 induces muscle relaxation via PKA- and Epac mediated inhibition of RhoA/Rho kinase pathway. *American Journal of Physiology; Gastrointestinal and Liver Physiology, 304*, G527–G535.

Reich, M., Deutschmann, K., Sommerfeld, A., Klindt, C., Kluge, S., Kubitz, R., Ullmer, C., Knoefel, W., Herebian, D., Mayatepek, E., Häussinger, D., & Keitel, V. (2016). TGR5 is essential for bile acid-dependent cholangiocyte proliferation in vivo and in vitro. *Gut, 65*, 487–501.

Röder, P., Wu, B., Liu, Y., & Han, W. (2016). Pancreatic regulation of glucose homeostasis. *Experimental and Molecular Medicine, 48*, e219.

Russell, D. (2003). The enzymes, regulation, and genetics of bile acid synthesis. *Annual Review of Biochemistry, 72*, 137–174.

Russell, D. (2009). Fifty years of advances in bile acid synthesis and metabolism. *Journal of Lipid Research, 50*, S120–S125.

Sirvent, A., Verhoeven, A., Jansen, H., Vladimir, K., Darteil, R., Hum, D., Jean-Charles, F., & Staels, B. (2004). Farnesoid X receptor represses hepatic lipase gene expression. *Journal of Lipid Research, 45*, 2110–2115.

Song, C., Hiipakka, R., & Liao, S. (2000). Selective activation of liver X receptor alpha by 6 alpha hydroxy bile acids and analogs. *Steroids, 2012*, 428139.

Taoufiq, H., Pols, T., Nomura, M., Maida, A., Watanabe, M., Auwerx, J., & Schoonjans, K. (2012). TGR5 potentiates GLP-1 secretion in response to anionic exchange resins. *Scientific Reports, 2*, 430.

Thomas, C., Gioiello, A., Noriega, L., Strehle, A., Oury, J., Rizzo, G., Macchiarulo, A., Yamamoto, H., Mataki, C., Pruzanski, M., Pellicciari, R., Auwerx, J., & Schoonjans, K. (2009). TGR5-mediated bile acid sensing controls glucose homeostasis. *Cellular Metabolsim, 10*, 167–177.

Tiwari, A., & Maiti, P. (2009). TGR5: An emerging bile acid G-protein-coupled receptor target for the potential treatment of metabolic disorders. *Drug Discovery Today, 14*, 523–530.

Trottier, J., Białek, A., Caron, P., Straka, R., Milkiewicz, P., & Barbier, O. (2011). Profiling Circulating and Urinary Bile Acids in Patients with Biliary Obstruction before and after Biliary Stenting. *PLOS One, 6*, e22094.

Ulrich, B., Trauner, M., Jansen, P., & Poupon, R. (2015). New paradigms in the treatment of hepatic cholestasis: From UDCA to FXR, PXR and beyond. *Journal of Hepatology, 62*, S25–S37.

Wang, H., Chen, J., Hollister, K., Sowers, L., & Forman, B. (1999). Endogenous bile acids are ligands for the nuclear receptor FXR/BAR. *Molecular and Cellular Biology, 3*, 543–553.

Wang, Y., Chen, W., Yu, D., Forman, B., & Huang, W. (2011). The g-protein coupled bile acid receptor, gpbar1 (tgr5), negatively regulates hepati inflammatory response through antagonizing nuclear factor kappa light-chain enhancer of activated b cells (nf-kappab) in mice. *Hepatology, 54*, 1421–1432.

Yutaka, S., Fukushima, M., & Yabe, D. (2010). GIP and GLP-1, the two incretin hormones: Similarities and differences. *Journal of Diabetes Investigation, 1*, 8–23.

Zhang, Y., Kast-Woelbern, H., & Edwards, P. (2003). Natural structural variants of the nuclear receptor farnesoid X receptor affect transcriptional activation. *Journal of Biological Chemistry, 278*, 101–110.

Zheng, X., Fengjie, H., Aihua, Z., Lei, S., Zhang, Y., Xie, G., Chen, T., Qu, C., Rajani, C., Dong, B., Li, D., & Jia, W. (2017). Bile acid is a significant host factor shaping the gut microbiome of diet-induced obese mice. *BMC Biology, 15*, 120.

Chapter 3
P2Y Receptor

Abstract Purinergic receptors (P2Ys) are rhodopsin-like GPCR (metabotropic receptors) that are activated by extracellular nucleotides (ATP, ADP, UTP and UDP). They stimulate a wide range of signaling pathways through activation of a variety of G proteins. P2Y family contains 8 members and they are grouped as P2Y1-like receptors (P2Y1, P2Y2, P2Y4, P2Y6 and P2Y11) and P2Y12-like receptors (P2Y12, P2Y13, P2Y14) based on their amino acid sequence homology and preference for ligands. These receptors are ubiquitously expressed in the biological system and they have unavoidable role in many physiological functions including neurotransmission, hormone secretion, blood vessel contraction, homeostasis and metabolism. As P2Y receptors can interact and/or heterodimerize with various GPCRs and other family receptors, they are now gaining much attention as a potential therapeutic target for various disorders including stroke, epilepsy and cancer. This chapter illustrates the classification, characteristic features and distribution of members of P2Y receptor family. In addition, the current knowledge of physiological and pathological functions of P2Y receptors are described.

Keywords Biochemical properties · Classification · Purinergic P2Y receptors · Physiological role · Pathological role

3.1 Introduction

Nucleotides are fundamental units of nucleic acid and free nucleotides are integral part of intracellular communication and metabolism by providing energy and acting as second messengers. The nucleotides and their metabolites such as adenosine triphosphate (ATP), adenosine diphosphate (ADP), adenosine triphosphate (ATP), adenosine diphosphate (ADP), uridine triphosphate (UTP), uridine diphosphate (UDP), and others are released from the cells through various routes. They functions as extracellular messengers under physiological conditions during which excitable and non-excitable cells normally releases nucleotides. However, the concentration and duration differs depending on the cell types. The extracellular nucleotides also act as a danger signal in pathological circumstances in which those

S. Rajagopal, M. Ponnusamy, *Metabotropic GPCRs: TGR5 and P2Y Receptors in Health and Diseases*, https://doi.org/10.1007/978-981-13-1571-8_3

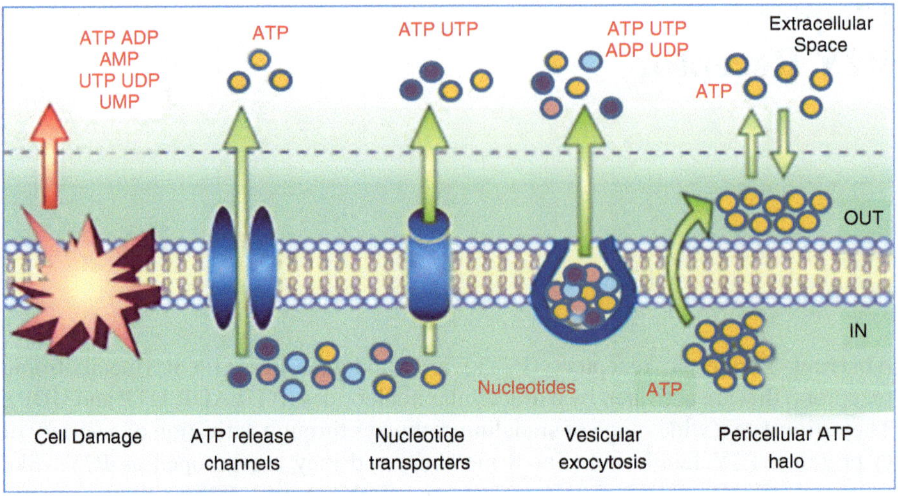

Fig. 3.1 Routes of nucleotide release from the cell

nucleotides are massively released from damaged or infected cells. They stimulate a unique family of cell surface receptors, which are called as 'purinergic receptors' and their signaling pathway is termed as 'purinergic signaling' (Burnstock 2006, 2012). Depends on their concentration and duration of activation, they execute different cellular functions. One classical example is ATP, which acts as a short-term signaling molecule in neurotransmission and secretion, while the tropic (long-term) effects of ATP is involved in the embryo development, regenerative response to injury, bone formation and immune system (Burnstock 2012) (Fig. 3.1).

The extracellular levels of nucleotides are controlled by a family of cell surface enzymes called as ectonucleotidases [ecto-nucleoside triphosphate diphosphohydrolase (E-NTPDase/CD39), ecto-nucleotide pyrophosphatase phosphodiesterases (E-NPPs) and ecto 5'nucleotidase (CD73)], which hydrolyze a variety of nucleotides to nucleosides. The typical members of cell surface E-NTPDase (NTPDase1, 2, 3, 8) sequentially converts nucleoside triphosphate (ATP, UTP) to nucleoside monophosphate (AMP, UMP) and the members of E-NPPs convert both nucleoside diphosphate (ADP, UDP) and triphosphate (ATP, UTP) to nucleoside monophosphate (AMP, UMP), while the ecto 5′nucleotidase hydrolyses only nucleoside monophosphate (AMP) to corresponding nucleoside (adenonise). Each member of this family has differential preferences for nucleotides (Fig. 3.2). As their activities modulate the levels of nucleotides and nucleosides in pericellular space, they play a vital role in tuning and controlling purinergic receptor signaling (Robson et al. 2006). In recent years, extracellular nucleotides are recognized as an important autocrine and/or paracrine agonists and they execute their actions in tissue microenvironment to modulate local cellular signaling. A growing number of evidence show that purinergic signaling has a crucial role in the pathophysiology of many disorders including kidney failure, cystic fibrosis, cancer, stroke, thrombosis, dry eye and brain disorders, which emphasize their clinical significance in treatment of various disorders.

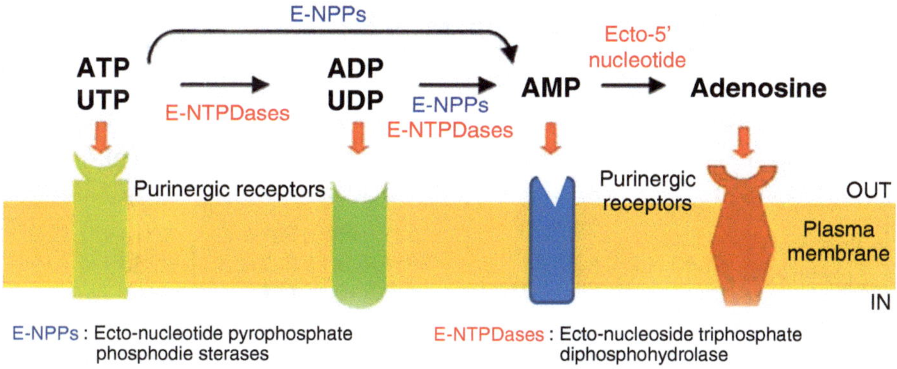

Fig. 3.2 Extracellular nucleotide metabolism and activation of purinergic receptors

3.2 Purinergic Receptors and Their Classification

The purinergic receptor superfamily contains 19 members and they are classified into two main subfamilies based on their affinity and preference for natural endogenous ligands (Burnstock 2012). The class I purine receptors (P1) are also called as adenosine receptors, which is only activated by purine adenosine. P1 subfamily has currently 4 members (A1, A2A, A2B, A3) and each is encoded by different genes in human. All these P1 receptors are typical rhodopsin-like GPCR (metabotropic) receptors and they are linked to specific G proteins (Gi/Go, Gs, Gs/Gq or Gi/Gq) and their downstream effectors (Tsuchiya and Nishizaki 2015). The class 2 purinergic receptors (P2) are activated by both purine and/or pyrimidine nucleotides (ADP/ATP, UDP/UTP). Thus, 'P2' denotes purine and pyrimidine, however, all P1 and P2 are collectively referred as purinergic receptors. P2 subfamily is further divided into two subtypes (P2X and P2Y) based on their functions. P2X receptors act as ligand-gated ionotropic nucleotide receptors, which have 7 members (P2X1–7). They are primarily involved in the fast excitatory neurotransmission and inflammatory response. But, P2Y members (8 subtypes) are slow acting metabotropic GPCRs that transduce signaling through G proteins (Burnstock, 2006, 2012) (Fig. 3.3).

3.3 P2Y Receptors

P2Y receptors are members of rhodopsin-like GPCRs that can activate or deactivate a variety of heterotrimeric/monomeric G proteins for the initiation of signal transduction. So far, 8 members of P2Y receptors (P2Y1, P2Y2, P2Y4, P2Y6, P2Y11, P2Y12, P2Y13, P2Y14) are identified and characterized in human (Erb and Weisman 2012). Given the fact that P2Y members have the capability to integrate a wide range of signaling through heteromerization with other members of GPCRs including P1 (A1R, A2AR) receptors and angiotensin AT1 receptors

Purinergic Receptors			
Family	Adenosine/P1 receptors (P1Rs)	P2 receptors (P2Rs)	
Subfamily	-	P2Y	P2X
Receptor Subtypes	A1, A2A, A2B, A3	P2Y1, P2Y2, P2Y4, P2Y6, P2Y11, P2Y12, P2Y13, P2Y14	P2X1, P2X2, P2X3, P2X4, P2X5, P2X6, P2X7
Structure	G protein-coupled receptors	G protein-coupled receptors	Ligand-gated ion channels
Ligands	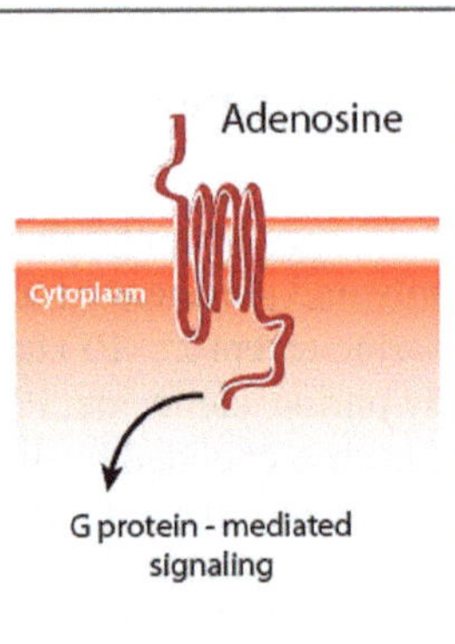	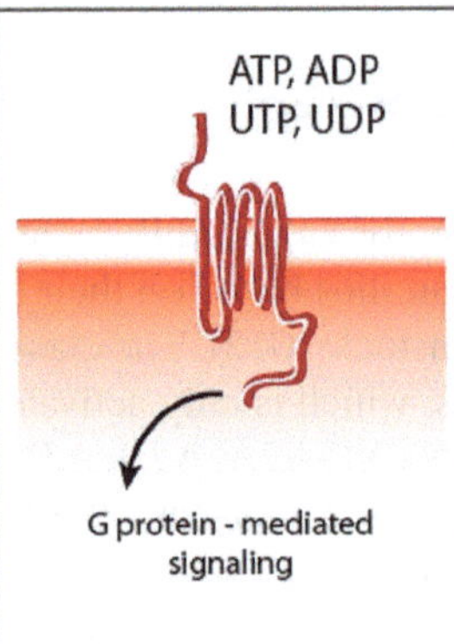	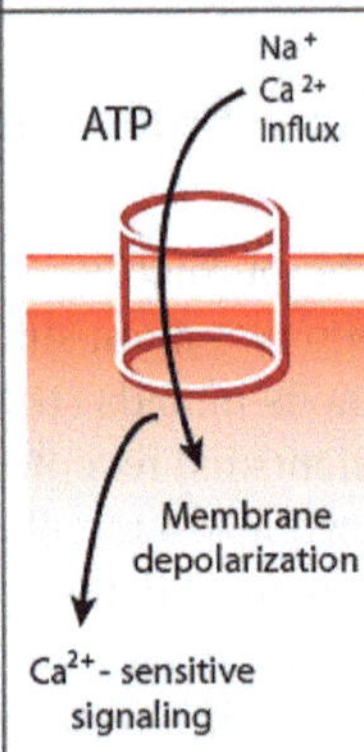

Fig. 3.3 Classification of purinergic receptors and their endogenous ligands

(Erb and Weisman 2012; Yoshioka et al. 2001; Suzuki et al. 2006; Nishimura et al. 2016), P2Y receptors have gained much attention among the other members of purinergic receptors.

3.3.1 P2Y Receptor Subtypes and Biochemical Properties

P2Y receptors have all typical features of GPCR including 7TM regions, three cytoplasmic loops and three extracellular loops. Unlike P2X receptors, P2Y members are activated by several nucleoside phosphates (ATP, ADP, UTP, UDP). The amino acid sequence of P2Y1, P2Y2, P2Y4, P2Y6 and P2Y11 share 28–52% homology, which are grouped into P2Y1-like receptors. Whereas P2Y12, P2Y13, and P2Y14 are grouped into P2Y12-like receptors due to 45–50% sequence homology (Nishimura et al. 2017). The structural studies found that all P2Y receptors, so far identified, have a common H-X-X-R/K motif in TM6, which is essential for agonist

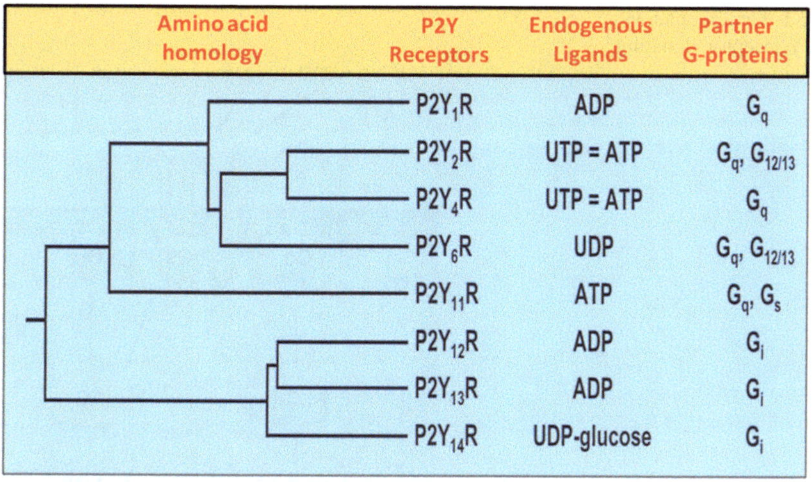

Fig. 3.4 Diversity of P2Y receptors in ligands and G protein Preference

binding and activity. The P2Y1- like receptors contain a common Y-Q/K-X-X-R motif in TM7, while PY12-like receptors share an additional motif (K-E-X-X-L) in TM7 (Abbracchio et al. 2003). Interestingly, an additional lysine (K) residue in extracellular loop is important for binding of nucleotides with P2Y12, P2Y13 and P2Y14 (Costanzi et al. 2004). These studies indicate that the molecular diversity of amino acid sequences in P2Y members causes variability in preferences, affinity and binding of ligands as well as heteromeric G-proteins. P2Y11 is only activated by ATP, while P2Y2 and P2Y4 are activated by ATP and UTP. The other members such as P2Y1, P2Y6, P2Y12, P2Y13 and P2Y14 are not responsive or less responsive to ATP stimulation. P2Y1, P2Y12 and P2Y13 have more affinity for ADP than other ligands. UDP required for P2Y6 activation, but P2Y14 is only stimulated by UDP-glucose. Similarly, P2Y specifically combine with one or more G-protein subtypes, which subsequently turn on defined intracellular signaling. P2Y1-like receptors mainly couple with Gq protein. Some of them can also bind with other G-proteins. For example, P2Y11 couples with Gs in addition to Gq, and P2Y2 and P2Y6 couple with G12/13 along with Gq. In contrast, P2Y12-like receptors such as P2Y12R, P2Y13R, and P2Y14R bind with Gi/o (Nishimura et al. 2017; Erb and Weisman 2012) (Fig. 3.4).

3.3.2 P2Y Receptor Signaling

The ligand activated P2Y1-like receptors (P2Y1, P2Y2, P2Y4, P2Y6 and P2Y11) results in coupling of Gq and subsequently stimulates PLC-β induced increase of cytoplasmic messengers such as inositol phosphate 3 (IP3) and/or diacyl glycerol (DAG), which promotes intracellular mobilization of Ca^{2+} (Suzuki et al. 2006;

Fig. 3.5 General pathway of P2Y receptor signaling

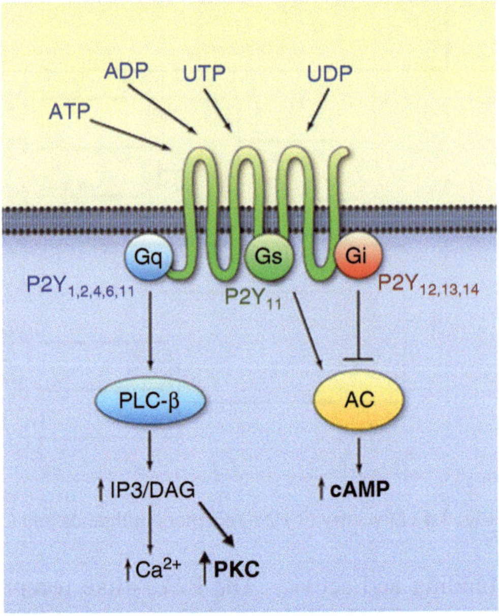

Tsuchiya and Nishizaki 2015). P2Y1 activates Gq dependent PLC-β signaling as well as monomeric G-protein (RhoA and Rac) signaling. P2Y2 and P2Y4 activate Gq/Go dependent PLC-β signaling. However, P2Y2 has the capability to induce Go dependent Rac activation as well as G12-Rho signaling. P2Y6 activates PLC-β and Rho signaling by coupling with Gq and G12/13, respectively. P2Y1-like receptors (P2Y12, P2Y13, and P2Y14) commonly targets an effector adenylyl cyclase (AC) in addition to PLC-β and/or RhoA. P2Y11 induces AC and PLC-β activity through coupling with Gs and Gq, respectively. On the other hand, P2Y12, P2Y13 and P2Y14 inhibit AC through Gi/o activation and positively modulate PLC-β and RhoA activity in P2Y12 and P2Y13 dependent signaling. But P2Y14- Gi/o pathway only activates PLC-β (Erb and Weisman 2012; Jacobson and Muller 2016) (Fig. 3.5).

In general, the desensitization of GPCR signaling is initiated by phosphorylation (Ser/Thr) of the intracellular domain of GPCR by GPCR kinases (GRK1-7), which recruits β-arrestin and promotes internalization and inactivation of GPCRs signal. P2Y receptor trafficking and signal termination also occurs in a classical way of GPCR desensitization (Abbracchio et al. 2003; Jacobson et al. 2015). In the recycling process, the binding of adaptor protein-2 to P2Y receptor – β-arrestin (β-Arr) complex is required for the internalization. However, the internalization component differs for some P2Y receptors. For example, P2Y2 and P2Y4 can interact with both β-Arr1 and β-Arr2, while P2Y12 interacts only with β-Arr2. The agonist ligand also alters the kinetics of binding. In the case of P2Y2 receptor, UTP stimulation induces strong interaction of P2Y2 receptor with both β-Arr1 and β-Arr2, but ATP stimulation only favours strong interaction with β-Arr1 and weak interaction with β-Arr2 (Nishimura et al. 2017). This variation in internalization also diverges the signaling after internalization. Another example for this event is P2Y12 internalisation in microglia cells and platelets. In human platelet cells, ADP mediated activation of P2Y12 stimulates small GTP-binding protein ADP ribosylation factor 6 (ARF6) activity, which is involved in P2Y12 as well as P2Y1 internalization. ARF6 promotes dynamin dependent endocytosis of P2Y12 receptors and it facilitates the resensitization of P2Y12 (Kanamarlapudi et al. 2012). On the other hand, P2Y12 internalization and resensitization in microglia is fully depending on β-Arr2, which activates ERK1/2 signaling after endocytosis (Lee et al. 2012). These findings reveal that the same receptor has distinct signaling pathways in different tissues that mainly depend on the target cellular function.

3.3.3 P2Y Receptor Distribution and Physiological Functions

P2Y receptors are ubiquitously expressed in all types of tissues. All P2Ys are highly expressed in brain, which implicates their physiolical role in neural cells. The majority of P2Ys also expressed in immune cells and their expression is associated with immune response and localized inflammatory response (Jacobson and Muller 2016). Some P2Y receptors such as P2Y14 have specialized functions in a tissue dependent manner. The placental expression of P2Y14 triggers innate mucosal immune response by activating neutrophil chemotaxis (Arase et al. 2009). P2Y1 and P2Y4 contribute to retinal development after birth by promoting muller cell differentiation (Wurm et al. 2009). P2Y13 take parts in cholesterol metabolism by promoting biliary secretion, cholesterol transport and HDL-cholesterol uptake by liver (Goffinet et al. 2014; Fabre et al. 2010). P2Y14 is involved in insulin secretion and glucose homeostasis regulation (Meister et al. 2014). The following table describes the distribution and major functions of P2Y receptors (Erb and Weisman 2012; Jacobson and Muller 2016) (Table 3.1).

3.3.4 P2Y Receptors in the Pathophysiology

3.3.4.1 P2Y in Cardiac Homeostasis and Pathophysiology

The endogenous nucleotides have impact on the cardiovascular system. The recent advancement of research in purinergic singalling demonstrates that P2 receptors significantly contribute to various cardiovascular disorders, including hypertension, vasospasm, congestive heart failure and cardiac damage during ischemic episodes. In cardiac tissue, the extracellular nucleotides, mainly ATP, is released from a variety of cells such as sympathetic nerves, endothelial cells and myocardial cells under various conditions including hypoxia, ischemic injury, mechanical stress and pressure overload. Thus, extracellular nucleotides have impact on physiological functions of the heart. The expression of several P2Y receptors including P2Y2 and

Table 3.1 P2Y receptor expression and their physiological functions

P2Y type	Distribution	Major physiological function
P2Y1	Brain (Cerebral cortex, Cerebellum, Hippocampus, Corpus callosum, Midbrain, Astrocytes), Placenta, Heart, Muscle, Prostate, Intestine, Platelets, Bone, Pancreas, Eye	Transmitter release, Changes in platelet shape and its aggregation, Angiogenesis, Bone resorption, Secretion of leptin from adipocytes, Mechanical and thermal nociception, Synaptic plasticity, Muller cell differentiation and retina development.
P2Y2	Pituitary, Heart, Blood, Vessels, Lung, Kidney, Placenta, Skeletal Muscle, Endocrine, Bone, Astrocytes	Transmitter release, Mucus escalation, Airway hydration, Epithelial K^+/Cl^- secretion, Cardiomyocyte Contraction, Recruitment of immune cells and phagocytosis, Thrombosis, Blood pressure and ocular pressure regulation, Mechanical and thermal nociception, Pancreatic function, Maintains water homeostasis and renal functions, Wound healing.
P2Y4	Brain, Placenta, Heart, Epithelium, Pancreas, Smooth muscle, Kidney, Intestine, Liver, Eye	Glial growth, Epithelial growth, Smooth muscle growth, Muller cell differentiation, Visual transmission, Auditory transduction, Intestinal secretion of K^+/Cl^-.
P2Y6	Kidney, Lung, Spleen, Thymus, Placenta, Heart, Bone, Smooth Muscle, Epithelium, Intestine	Lymphocytic Maturation, Cardiomyocyte Contraction, Bone resorption, Microglial phagocytosis, Epithelial Cl− secretion.
P2Y11	Brain, Spleen, Placenta, Intestine, Smooth muscle, Granulocytes	Granulocyte differentiation, Cardiomyocyte Contraction, Neutrophil survival, Negative regulation of TLR signaling, Pancreatic endocrine and exocrine secretion.
P2Y12	Brain, Platelets	Haemostasis, Blood clotting, Microglial activation and migration, Activation of Dendritic cells and macropinocytosis, Peripheral anti-nociception.
P2Y13	Brain, Spleen, Bone marrow, Lymph nodes	Transmitter release, Differentiation of bone marrow cells, Bone formation, Cholesterol metabolism regulation, Hepatic uptake of HDL, Inhibition of ATP release from RBCs, Peripheral anti-nociception.
P2Y14	Pituitary, Brain, Placenta, Skeletal muscle, Adipose tissue, Intestine, Spleen, Lung, Stomach, Heart, Peripheral blood Mononuclear cells	Insulin secretion, Neuroimmune function, Innate immunity to placenta, Immune response in epithelial cells, Stomach contractility and Gastric function, Peripheral anti-nociception.

P2Y6 have beneficial role in the heart by improving contractile function (Wihlborg et al. 2006; Cohen et al. 2011), promoting angiogenesis, reducing cardiomyocyte cell death (Hochhauser et al. 2013), preventing pathological hypertrophy (Nishida et al. 2011) and reducing fibrosis after myocardial injury. However, one study shows that P2Y6 dependent Gα12/13-RhoA signalling in cardiomyocytes stimulate profibrotic gene expression and fibrotic response in a pressure overload animal model (Nishida et al. 2008). In addition, the activated P2Y6 interacts with angiotensin receptor (AT1R) and forms a stable herterodimer, which enhances G-protein dependent vascular hypertrophy along with suppression of β-Arr mediated AT1R internalization (Nishimura et al. 2016). P2Y4 also induces pathological hypertrophy and remodeling by upregulating endothelin-1 expression, inflammatory response and neutrophil infiltration (Horckmans et al. 2015). Currently, the influence of purinergic signaling including P2Y receptors in the pathogenesis of cardiac fibrosis is not completely understood. Available reports indicate that extracellular ATP and other nucleotides can induce expression of profibrotic genes involved in fibroblast activation (conversion of myofibroblast) and proliferation. A mechanistic study found that ATP released from cardiomyocytes via connexin-43 during ischemia fuels fibroblast activation and proliferation by activation of P2Y2 receptor. Another research group found that ATP released from cardiomyocytes through pannexin-1 under the pathological conditions upregulates P2Y6 expression and activation in cardiomyocytes in autocrine manner. This leads to upregulation of expression and secretion of profibrotic factors and induces transformation of fibroblast to a fibrotic phenotype (Novitskaya et al. 2016).

3.3.4.2 P2Y in Vascular Remodeling, Atherosclerosis and Hypertension

Among the three layers of blood vessel wall (intima, media and adventitia), the intact media (a contractile layer) is vital for controlling the blood pressure that is predominantly composed of vascular smooth muscle cells (VSMC). The elasticity of media layer depends on VSMCs and that determines the vascular tone. P2Y receptors activation through nucleotide release from endothelial cells triggers the vascular tone regulatory processes. It is well defined that ATP/ADP dependent activation of P2Y signaling can mediate vasoconstriction in both human and animals (Mitchell et al. 2012). The vascular endothelial cell expresses Gq- coupled P2Y1, P2Y2 and P2Y6. The experimental studies using mice model and cell lines found that Gq/11-coupled P2Y2 is crucial for the sheer stress induced generation of endothelial NO synthatase and vasodilatation. The defective or lack of P2Y2 expression causes transformation of VSMC to a hypertensive phenotype, which subsequently leads to vascular tone, increased blood pressure and hypertension (Wihlborg et al. 2003, 2004). In contrast, UDP dependent P2Y receptor (P2Y6) signaling promotes arterial myogenic tone, vasoconstriction and heart failure and this pathological effect is triggered by P2Y6 mediated by Gq/11 dependent Ca^{2+} release and G12/13-RhoA signaling (Kauffenstein et al. 2016). The vascular injury triggers vascular remodeling and atherosclerosis by a series of changes in vascular wall that includes stimulation of VSMC and endothelial cell proliferation, migration and extracellular matrix modifications. The release of nucleotides from the injury sites and activation of P2Y receptor system of VSMC and endothelial cells can be one of the mechanisms of changes in the vascular wall. In vitro studies confirmed that ADP induced activation of P2Y1 and P2Y13 can promote endothelial cell proliferation and ATP and UTP promote VSMC proliferation through PI3K and ERK1/2, which indicates the link between G coupled P2Y receptors and pathological vascular remodeling

(Nishimura et al. 2017). The inflammatory response leads to invasion and attachment of immune cells to the injury sites, which consequently causes atherosclerotic plaque formation due to aggregation of cells. The endothelial P2Y2R signaling also promotes inflammatory response by upregulation of pro-inflammatory cytokines (VCAM1) and macrophage invasion at the site of injury. P2Y6 can also induce VCAM1 expression and macrophage activation under the condition of high fat in circulation, which ultimately develop atheroscelerotic lesions (Nishimura et al. 2016, 2017; Zerr et al. 2011). P2Y1, P2Y2, P2Y6 and P2Y12 have atheroscleorotic activity (Nishimura et al. 2016, 2017; West et al. 2014), while the expression of P2Y13 opposes this effect (Lichtenstein et al. 2015). In addition, P2Y1 and P2Y12 closely associated with the activation of platelets and thrombus development, which leads to risk for myocardial infarction and heart failure. P2Y1-Gq and P2Y12-Gα12 inhibit cAMP production by blocking adenylyl cyclase, which is a major suppressor of platelet aggregation (Andre et al. 2003; Eckly et al. 2001). These findings suggest that the balanced activities of different P2Y receptors are indispensable for combating the pathological vascular alterations and atherosclerosis.

3.3.4.3 P2Y in the Liver Homeostasis and Disorders

Several members of purinergic receptors participate in the normal liver physiological process including bile secretion, glycogen and lipid metabolism. The mechanical stress or any injury causes the release of ATP and other nucleotides from hepatocytes and Kupffer cells. The extracellular ATP contributes to the liver regenerative response by influencing hepatocyte proliferation through various purinergic receptors including P2Y receptor activation (Burnstock et al. 2014). In vivo studies found that P2Y2 can promote liver regeneration by accelerating hepatocyte proliferation through ERK1/2 and JNK dependent pathways. The extracellular concentration of ATP less <10 nM is enough to potentiate the growth factors induced hepatocyte proliferation and tissue regeneration after partial hepactomy (Thevananther et al. 2004). Another study using P2Y2 knockout animal model confirmed the role of P2Y2 in the hepatocyte proliferation and liver regeneration. However, P2Y2 activates ERK1/2, which is a mitogenic signal promotes cell cycle progression of hepatocytes (Tackett et al. 2014). Another research group observed that the extracellular ATP induces P2Y2 expression in acute hepatitis condition and that promotes liver infiltration by activating neutrophils, which causes excessive hepatocyte cell death. The administration of suramin, a non-specific P2Y inhibitor blocks this event (Ayata et al. 2012). P2Y receptors including P2Y13 are involved in the activation of resident macrophage (Kupffer cells) and release of pro-inflammatory cytokines (IL6 and TNFα), which causes liver inflammation and liver injury (Ishimaru et al. 2014). P2Y12 is required to induce sepsis induced systemic inflammation and multiorgan damage including liver that mainly occurs through P2Y12 dependent activation of neutrophil migration. Similarly, ATP-induced activation of P2Y2 and P2Y11 accelerates the proliferation and migration of hepatocellular carcinoma cells (Khalid et al. 2017; Xie et al. 2014), which shows its oncogenic

Table 3.2 Expression and functions of P2Y receptors in the liver

Cell type of the liver	P2Y Expression	Function
Hepatocytes	$P2Y_1$, $P2Y_2$, $P2Y_4$, $P2Y_6$, $P2Y_{13}$	Regulation of glycogenolysis and Glycogen metabolism, HDL-cholesterol uptake, insulin resistance
Cholangiocytes	$P2Y_1$, $P2Y_2$, $P2Y_4$, $P2Y_6$	Secretion of Cl- and bicaronate, salvage of nucleotides, canalicular contraction, interaction with hepatocytes, Resorption of adenosine from bile
Endothelial cells	$P2Y_1$, $P2Y_2$, $P2Y_6$	NO release, secretion of prostaglandins E2
Vascular smooth muscle cells	$P2Y_1$, $P2Y_2$, $P2Y_6$	Portal vein contraction
Hepatic stellate cells	$P2Y_2$, $P2Y_4$, $P2Y_6$	prostaglandin F2 and D2 secretion, cell contraction
Kupffer cell/macrophages	$P2Y_1$, $P2Y_2$, $P2Y_4$, $P2Y_6$	Killing of intracellular pathogens, secretion of prostaglandin E2, interleukin-6
Liver-associated lymphocytes: Natural Killer cells and Natural killer T cells, T cells, B cells	$P2Y_1$, $P2Y_2$, $P2Y_4$, $P2Y_6$	protects from concanavalin A-mediated hepatitis
Neutrophils	$P2Y_2$	Chemotaxis

role in hepatic cancer. Currently, few reports are available to demonstrate the role of P2Y in liver pathology. Nevertheless, the majority of those studies clearly indicate that P2Y receptors have primarily contribute to inflammatory response by triggering the production and release of pro-inflammatory cytokines. Some of the reports also highlight the positive role of P2Y receptors, especially P2Y2 favours hepatocytes proliferation, in promoting liver regeneration (Thevananther et al. 2004; Tackett et al. 2014) (Table 3.2).

3.3.4.4 Role of P2Y in Neurological Disorders

The cells of central nerves system express almost all subtypes of P2Y receptors in which they have several vital physiological functions. P2Y receptors play a significant role in the regulation and maintenance of blood brain barrier. P2Y members have distinct functions, both neuroprotection and detrimental effects, including neurotransmission, neural cell growth and differentiation, cell death induction. The neural precursor cells (NPC) are multipotent stem cells of CNS, which have the potency to regenerate and produce the major cells (neurons, oligodendrocytes and astrocytes) of the nerves system. Among the members of P2Y family, P2Y1 and P2Y2 are expressed by NPCs and these two receptors promote nucleotide-stimulated Ca^{2+} oscillation in those cells, which is a vital process for the NPC proliferation during embryo development. In addition, P2Y1 and P2Y2 regulate migration and differentiation of NPCs. These findings reveal the contribution of P2Y receptors in

Table 3.3 Functions of P2Y receptors in the central nerves system

P2Y receptors	Functions
P2Y1	Induces neural precursor cell proliferation, migration and differentiation, Promotes cytokine/chemokine-induced damage following ischemia, Promotes axonal elongation and modulation of pain sensation, Promotes anxiolytic behavior.
P2Y2	Neural precursor cell proliferation, migration and differentiation, Increases the migration of glial cells by interacting with integrin, Promotes glial cell proliferation, Promotes non- amyloidogenic APP processing in neurons, Promotes the uptake and degradation of neurotoxic forms of $A\beta_{1-42}$ by microglial cell.
$P2Y_4$	Inhibits presynaptic glutamate release, Modulates blood-brain barrier function, Inhibits K^+ currents in rat myocytes.
P2Y6	Increases microglial phagocytic activity, Regulates CNS repair mechanisms.
P2Y11	Blocks or delays neutrophil apoptosis induced by pathogen- or inflammation, Negatively regulates TLR signaling and modulates cytokine secretion.
P2Y12	Causes platelet aggregation and leads to ischemia and stroke, Activates and regulates glial cell migration, Promotes glial cell proliferation.
P2Y13	Increases glycine transport in the synaptic cleft, Promotes cell survival through activation of PI3K/Akt-signaling.
P2Y14	Promotes chemokine and cytokine production and modulates inflammatory responses

neurogenesis and CNS repair following injury (Peter Illes et al. 2013). The deposition of Amyloid β-protein ($A\beta_{1-42}$) from dead cells in the Alzheimer's brain leads to worsening of disease. The rapid activation of P2Y2 in microglial cells by nucleotides released from dead cells promotes the phagocytic activity of microglial cells and enhances the clearance of $A\beta_{1-42}$. P2Y2 activates α(V) integrins/Src/Rac signaling pathway to induce microglial phagocytosis (Kim et al. 2012). In addition, P2Y2 interacts with nerve growth factor (NGF)/tyrosine receptor kinase A (TrkA) and this interaction favors neuronal differentiation by enhancing neurite formation, which is a central event in neurodevelopment and regeneration after injury (Arthur et al. 2005). As P2Y are involved in the generation of pro-inflammatory factors and neuroinflammation, they are closely associated many neurological disorders including epilepsy, seizure, ischemic stroke and neuropathic pain. The influence of P2Y receptors on the repair mechanisms of CNS that elicits following inflammation, tissue damage and pathological disorders, which represents the importance of P2Y receptors as a target for the treatment of neurodegenerative disorders (Weisman et al. 2012; Jacobson and Muller 2016). The known functions of P2Y subtypes in CNS are given in Table 3.3.

3.3.4.5 Role of P2Y in Other Pathological Disorders

As proinflammatory reaction is one of the major mechanisms of extracellular nucleotides mediated tissue damage, purinergic signaling has been linked to the pathophysiology of many disorders including asthma. A recent finding reveals that acute and chronic allergic reactions of lung tissue lead to upregulation of P2Y6 receptor expression in airway epithelial cells, which induces Th2 cytokine production and hyperresponsiveness in bronchial tissue. The genetic or pharmacological blockage of P2Y6 attenuates this pathological event in lungs (Vieira et al. 2011). In contrast, ATP/UTP dependent activation of P2Y receptors such as P2Y2 and P2Y6 stimulate human lung adenocarcinoma by promoting cell migration and proliferation through a mechanism related to intracellular Ca^{2+} mobilization and NFkB signaling (Schafer et al. 2003; Wang et al. 2017). In the kidney, the expression of P2Y12 impairs the autoregulatory ability of renal arterial pressure following AngII-mediated hypertension, which leads to thickening of preglomerular vessels, tubular damage and interstitial fibrosis (Osmond et al. 2014). During acute kidney injury, the activation of P2Y receptors, but not other class of nucleotide receptors (P2X and P1), protects further damage by inhibiting NFkB signaling. The exogenous administration of ATP and P2Y agonist (ATP-gamma-S) promotes P2Y mediated protection against ischemic damage (Lee and Han 2005). In chronic renal disease, P2Y2 expression is required for the compensatory hypertrophy and blood pressure maintenance (Potthoff et al. 2013).

A high expression of P2Y14 (GPR105) in collecting duct intercalated cells (ICs) of kidney elicits sterile inflammation by recruiting neutrophils. Since uncontrolled inflammation during the injury and spontaneous inflammatory response without any infection causes renal damage and deterioration of functions, the role of P2Y14 in sterile inflammation is a novel finding for the therapeutic strategy to prevent acute kidney injury (Azroyan et al. 2015). Collectively, these findings reveal that P2Y receptors have a wide range of physiological as well as pathological functions.

3.4 Summary

The extracellular nucleotide activated P2Y receptors, the metabotropic GPCRs, are gaining attention in recent years as a therapeutic target for many disorders due to the expanding list of their physiological as well as pathological role. In this chapter, we have described the well defined signaling pathways, physiological functions and pathological disorders associated with P2Y receptors. However, our understanding about the biological functions of P2Y is still infancy. As P2Y receptors come into limelight over the recent decade, we could expect more effective therapeutic agents targeting P2Y in the future to combat various pathological disorders.

References

Abbracchio, M. P., Boeynaems, J. M., Barnard, E. A., Boyer, J. L., Kennedy, C., Miras-Portugal, M. T., King, B. F., Gachet, C., Jacobson, K. A., Weisman, G. A., & Burnstock, G. (2003). Characterization of the UDP-glucose receptor (re-named here the P2Y14 receptor) adds diversity to the P2Y receptor family. *Trends in Pharmacological Science, 24*, 52–55.

Andre, P., Delaney, S. M., LaRocca, T., Vincent, D., DeGuzman, F., Jurek, M., Koller, B., Phillips, D. R., & Conley, P. B. (2003). P2Y12 regulates platelet adhesion/activation, thrombus growth, and thrombus stability in injured arteries. *Journal of Clinical Investigation, 112*, 398–406.

Arase, T., Uchida, H., Kajitani, T., Ono, M., Tamaki, K., Oda, H., Nishikawa, S., Kagami, M., Nagashima, T., Masuda, H., Asada, H., Yoshimura, Y., & Maruyama, T. (2009). The UDP-glucose receptor P2RY14 triggers innate mucosal immunity in the female reproductive tract by inducing IL-8. *Journal of Immunology, 182*, 7074–7084.

Arthur, D. B., Akassoglou, K., & Insel, P. A. (2005). P2Y2 receptor activates nerve growth factor/TrkA signaling to enhance neuronal differentiation. *Proceedings of the National Academy of Sciences, USA, 102*, 19138–19143.

Ayata, C. K., Ganal, S. C., Hockenjos, B., Willim, K., Vieira, R. P., Grimm, M., Robaye, B., Boeynaems, J. M., Di Virgilio, F., Pellegatti, P., Diefenbach, A., Idzko, M., & Hasselblatt, P. (2012). Purinergic P2Y(2) receptors promote neutrophil infiltration and hepatocyte death in mice with acute liver injury. *Gastroenterology, 143*(1620–1629), e1624.

Azroyan, A., Cortez-Retamozo, V., Bouley, R., Liberman, R., Ruan, Y. C., Kiselev, E., Jacobson, K. A., Pittet, M. J., Brown, D., & Breton, S. (2015). Renal intercalated cells sense and mediate inflammation via the P2Y14 receptor. *PLoS One, 10*, e0121419.

Burnstock, G. (2006). Purinergic signalling. *British Journal of Pharmacology, 147*(Suppl 1), S172–S181.

Burnstock, G. (2012). Purinergic signalling: Its unpopular beginning, its acceptance and its exciting future. *BioEssays, 34*, 218–225.

Burnstock, G., Vaughn, B., & Robson, S. C. (2014). Purinergic signalling in the liver in health and disease. *Purinergic Signalling, 10*, 51–70.

Cohen, R., Shainberg, A., Hochhauser, E., Cheporko, Y., Tobar, A., Birk, E., Pinhas, L., Leipziger, J., Don, J., & Porat, E. (2011). UTP reduces infarct size and improves mice heart function after myocardial infarct via P2Y2 receptor. *Biochemical Pharmacology, 82*, 1126–1133.

Costanzi, S., Mamedova, L., Gao, Z. G., & Jacobson, K. A. (2004). Architecture of P2Y nucleotide receptors: structural comparison based on sequence analysis, mutagenesis, and homology modeling. *Journal of Medicnal Chemistry, 47*, 5393–5404.

Eckly, A., Gendrault, J. L., Hechler, B., Cazenave, J. P., & Gachet, C. (2001). Differential involvement of the P2Y1 and P2YT receptors in the morphological changes of platelet aggregation. *Thrombosis and Haemostasis, 85*, 694–701.

Erb, L., & Weisman, G. A. (2012). Coupling of P2Y receptors to G proteins and other signaling pathways. *Wiley Interdisciplinary Reviews-Developmental Biology, 1*, 789–803.

Fabre, A. C., Malaval, C., Ben Addi, A., Verdier, C., Pons, V., Serhan, N., Lichtenstein, L., Combes, G., Huby, T., Briand, F., Collet, X., Nijstad, N., Tietge, U. J., Robaye, B., Perret, B., Boeynaems, J. M., & Martinez, L. O. (2010). P2Y13 receptor is critical for reverse cholesterol transport. *Hepatology, 52*, 1477–1483.

Goffinet, M., Tardy, C., Boubekeur, N., Cholez, G., Bluteau, A., Oniciu, D. C., Lalwani, N. D., Dasseux, J. L., Barbaras, R., & Baron, R. (2014). P2Y13 receptor regulates HDL metabolism and atherosclerosis in vivo. *PLoS One, 9*, e95807.

Hochhauser, E., Cohen, R., Waldman, M., Maksin, A., Isak, A., Aravot, D., Jayasekara, P. S., Muller, C. E., Jacobson, K. A., & Shainberg, A. (2013). P2Y2 receptor agonist with enhanced stability protects the heart from ischemic damage in vitro and in vivo. *Purinergic Signalling, 9*, 633–642.

Horckmans, M., Esfahani, H., Beauloye, C., Clouet, S., di Pietrantonio, L., Robaye, B., Balligand, J. L., Boeynaems, J. M., Dessy, C., & Communi, D. (2015). Loss of mouse P2Y4 nucleotide receptor protects against myocardial infarction through endothelin-1 downregulation. *Journal of Immunology, 194*, 1874–1881.

Ishimaru, M., Yusuke, N., Tsukimoto, M., Harada, H., Takenouchi, T., Kitani, H., & Kojima, S. (2014). Purinergic signaling via P2Y receptors up-mediates IL-6 production by liver macrophages/Kupffer cells. *Journal of Toxicological Sciences, 39*, 413–423.

Jacobson, K. A., & Muller, C. E. (2016). Medicinal chemistry of adenosine, P2Y and P2X receptors. *Neuropharmacology, 104*, 31–49.

Jacobson, K. A., Paoletta, S., Katritch, V., Wu, B., Gao, Z. G., Zhao, Q., Stevens, R. C., & Kiselev, E. (2015). Nucleotides acting at P2Y receptors: Connecting structure and function. *Molecular Pharmacology, 88*, 220–230.

Kanamarlapudi, V., Owens, S. E., Saha, K., Pope, R. J., & Mundell, S. J. (2012). ARF6-dependent regulation of P2Y receptor traffic and function in human platelets. *PLoS One, 7*, e43532.

Kauffenstein, G., Tamareille, S., Prunier, F., Roy, C., Ayer, A., Toutain, B., Billaud, M., Isakson, B. E., Grimaud, L., Loufrani, L., Rousseau, P., Abraham, P., Procaccio, V., Monyer, H., de Wit, C., Boeynaems, J. M., Robaye, B., Kwak, B. R., & Henrion, D. (2016). Central role of P2Y6 UDP receptor in arteriolar myogenic tone. *Arteriosclerosis, Thrombosis and Vascular Biology, 36*, 1598–1606.

Khalid, M., Brisson, L., Tariq, M., Hao, Y., Guibon, R., Fromont, G., Mortadza, S. A. S., Mousawi, F., Manzoor, S., Roger, S., & Jiang, L. H. (2017). Carcinoma-specific expression of P2Y11 receptor and its contribution in ATP-induced purinergic signalling and cell migration in human hepatocellular carcinoma cells. *Oncotarget, 8*, 37278–37290.

Kim, H. J., Ajit, D., Peterson, T. S., Wang, Y., Camden, J. M., Gibson Wood, W., Sun, G. Y., Erb, L., Petris, M., & Weisman, G. A. (2012). Nucleotides released from Abeta(1)(-)(4)(2) -treated microglial cells increase cell migration and Abeta(1)(-)(4)(2) uptake through P2Y(2) receptor activation. *Journal of Neurochemistry, 121*, 228–238.

Lee, Y. J., & Han, H. J. (2005). Effect of adenosine triphosphate in renal ischemic injury: involvement of NF-kappaB. *Journal of Cellular Physiology, 204*, 792–799.

Lee, S. H., Hollingsworth, R., Kwon, H. Y., Lee, N., & Chung, C. Y. (2012). beta-arrestin 2-dependent activation of ERK1/2 is required for ADP-induced paxillin phosphorylation at Ser(83) and microglia chemotaxis. *Glia, 60*, 1366–1377.

Lichtenstein, L., Serhan, N., Espinosa-Delgado, S., Fabre, A., Annema, W., Tietge, U. J., Robaye, B., Boeynaems, J. M., Laffargue, M., Perret, B., & Martinez, L. O. (2015). Increased atherosclerosis in P2Y13/apolipoprotein E double-knockout mice: contribution of P2Y13 to reverse cholesterol transport. *Cardiovascular Research, 106*, 314–323.

Meister, J., Le Duc, D., Ricken, A., Burkhardt, R., Thiery, J., Pfannkuche, H., Polte, T., Grosse, J., Schoneberg, T., & Schulz, A. (2014). The G protein-coupled receptor P2Y14 influences insulin release and smooth muscle function in mice. *Journal of Biological Chemistry, 289*, 23353–23366.

Mitchell, C., Syed, N. I., Tengah, A., Gurney, A. M., & Kennedy, C. (2012). Identification of contractile P2Y1, P2Y6, and P2Y12 receptors in rat intrapulmonary artery using selective ligands. *Journal of Pharmacology and Experimental Therapeutics, 343*, 755–762.

Nishida, M., Sato, Y., Uemura, A., Narita, Y., Tozaki-Saitoh, H., Nakaya, M., Ide, T., Suzuki, K., Inoue, K., Nagao, T., & Kurose, H. (2008). P2Y6 receptor-Galpha12/13 signalling in cardiomyocytes triggers pressure overload-induced cardiac fibrosis. *EMBO Journal, 27*, 3104–3115.

Nishida, M., Ogushi, M., Suda, R., Toyotaka, M., Saiki, S., Kitajima, N., Nakaya, M., Kim, K. M., Ide, T., Sato, Y., Inoue, K., & Kurose, H. (2011). Heterologous down-regulation of angiotensin type 1 receptors by purinergic P2Y2 receptor stimulation through S-nitrosylation of NF-kappaB. *Proceedings of the Natational Academy of Sciences, USA, 108*, 6662–6667.

Nishimura, A., Sunggip, C., Tozaki-Saitoh, H., Shimauchi, T., Numaga-Tomita, T., Hirano, K., Ide, T., Boeynaems, J. M., Kurose, H., Tsuda, M., Robaye, B., Inoue, K., & Nishida, M. (2016). Purinergic P2Y6 receptors heterodimerize with angiotensin AT1 receptors to promote angiotensin II-induced hypertension. *Science Signaling, 9*, ra7.

Nishimura, A., Sunggip, C., Oda, S., Numaga-Tomita, T., Tsuda, M., & Nishida, M. (2017). Purinergic P2Y receptors: Molecular diversity and implications for treatment of cardiovascular diseases. *Pharmacology and Therapeutics, 180*, 113–128.

Novitskaya, T., Chepurko, E., Covarrubias, R., Novitskiy, S., Ryzhov, S. V., Feoktistov, I., & Gumina, R. J. (2016). Extracellular nucleotide regulation and signaling in cardiac fibrosis. *Journal of Molecular and Cellular Cardiology, 93*, 47–56.

Osmond, D. A., Zhang, S., Pollock, J. S., Yamamoto, T., De Miguel, C., & Inscho, E. W. (2014). Clopidogrel preserves whole kidney autoregulatory behavior in ANG II-induced hypertension. *American Journal of Physiology Renal Physiology, 306*, F619–F628.

Peter Illes, P., Messemer, N., & Rubini, P. (2013). P2Y receptors in neurogenesis. *WIREs Membrane Transport and Signaling, 2*, 43–48.

Potthoff, S. A., Stegbauer, J., Becker, J., Wagenhaeuser, P. J., Duvnjak, B., Rump, L. C., & Vonend, O. (2013). P2Y2 receptor deficiency aggravates chronic kidney disease progression. *Frontiers in Physiology, 4*, 234.

Robson, S. C., Sevigny, J., & Zimmermann, H. (2006). The E-NTPDase family of ectonucleotidases: Structure function relationships and pathophysiological significance. *Purinergic Signalling, 2*, 409–430.

Schafer, R., Sedehizade, F., Welte, T., & Reiser, G. (2003). ATP- and UTP-activated P2Y receptors differently regulate proliferation of human lung epithelial tumor cells. *American Journal of Physiology Lung Cellular and Molecular Physiology, 285*, L376–L385.

Suzuki, T., Namba, K., Tsuga, H., & Nakata, H. (2006). Regulation of pharmacology by hetero-oligomerization between A1 adenosine receptor and P2Y2 receptor. *Biochemistry and Biophysics Research Communication, 351*, 559–565.

Tackett, B. C., Sun, H., Mei, Y., Maynard, J. P., Cheruvu, S., Mani, A., Hernandez-Garcia, A., Vigneswaran, N., Karpen, S. J., & Thevananther, S. (2014). P2Y2 purinergic receptor activation is essential for efficient hepatocyte proliferation in response to partial hepatectomy. *American Journal of Physiology Gastrointestinal and Liver Physiology, 307*, G1073–G1087.

Thevananther, S., Sun, H., Li, D., Arjunan, V., Awad, S. S., Wyllie, S., Zimmerman, T. L., Goss, J. A., & Karpen, S. J. (2004). Extracellular ATP activates c-jun N-terminal kinase signaling and cell cycle progression in hepatocytes. *Hepatology, 39*, 393–402.

Tsuchiya, A., & Nishizaki, T. (2015). Anticancer effect of adenosine on gastric cancer via diverse signaling pathways. *World Journal of Gastroenterology, 21*, 10931–10935.

Vieira, R. P., Muller, T., Grimm, M., von Gernler, V., Vetter, B., Durk, T., Cicko, S., Ayata, C. K., Sorichter, S., Robaye, B., Zeiser, R., Ferrari, D., Kirschbaum, A., Zissel, G., Virchow, J. C., Boeynaems, J. M., & Idzko, M. (2011). Purinergic receptor type 6 contributes to airway inflammation and remodeling in experimental allergic airway inflammation. *American Journal of Respiratory and Critical Care Medicine, 184*, 215–223.

Wang, X., Li, L., Guan, R., Zhu, D., Song, N., & Shen, L. (2017). Emodin inhibits ATP-induced proliferation and migration by suppressing P2Y receptors in human lung adenocarcinoma cells. *Cellular Physiology and Biochemistry, 44*, 1337–1351.

Weisman, G. A., Woods, L. T., Erb, L., & Seye, C. I. (2012). P2Y receptors in the mammalian nervous system: pharmacology, ligands and therapeutic potential. *CNS and Neurological Disorders Drug Targets, 11*, 722–738.

West, L. E., Steiner, T., Judge, H. M., Francis, S. E., & Storey, R. F. (2014). Vessel wall, not platelet, P2Y12 potentiates early atherogenesis. *Cardiovascular Research, 102*, 429–435.

Wihlborg, A. K., Malmsjo, M., Eyjolfsson, A., Gustafsson, R., Jacobson, K., & Erlinge, D. (2003). Extracellular nucleotides induce vasodilatation in human arteries via prostaglandins, nitric oxide and endothelium-derived hyperpolarising factor. *British Journal of Pharmacology, 138*, 1451–1458.

Wihlborg, A. K., Wang, L., Braun, O. O., Eyjolfsson, A., Gustafsson, R., Gudbjartsson, T., & Erlinge, D. (2004). ADP receptor P2Y12 is expressed in vascular smooth muscle cells and stimulates contraction in human blood vessels. *Arteriosclerosis, Thrombosis and Vascular Biology, 24*, 1810–1815.

Wihlborg, A. K., Balogh, J., Wang, L., Borna, C., Dou, Y., Joshi, B. V., Lazarowski, E., Jacobson, K. A., Arner, A., & Erlinge, D. (2006). Positive inotropic effects by uridine triphosphate (UTP) and uridine diphosphate (UDP) via P2Y2 and P2Y6 receptors on cardiomyocytes and release of UTP in man during myocardial infarction. *Circulation Research, 98*, 970–976.

Wurm, A., Erdmann, I., Bringmann, A., Reichenbach, A., & Pannicke, T. (2009). Expression and function of P2Y receptors on muller cells of the postnatal rat retina. *Glia, 57*, 1680–1690.

Xie, R., Xu, J., Wen, G., Jin, H., Liu, X., Yang, Y., Ji, B., Jiang, Y., Song, P., Dong, H., & Tuo, B. (2014). The P2Y2 nucleotide receptor mediates the proliferation and migration of human hepatocellular carcinoma cells induced by ATP. *Journal of Biological Chemistry, 289*, 19137–19149.

Yoshioka, K., Saitoh, O., & Nakata, H. (2001). Heteromeric association creates a P2Y-like adenosine receptor. *Proceedings of the National Academy of Sciences, USA, 98*, 7617–7622.

Zerr, M., Hechler, B., Freund, M., Magnenat, S., Lanois, I., Cazenave, J. P., Leon, C., & Gachet, C. (2011). Major contribution of the P2Y(1)receptor in purinergic regulation of TNFalpha-induced vascular inflammation. *Circulation, 123*, 2404–2413.

Chapter 4
Therapeutically Targeting TGR5 and P2Y Receptors

Abstract A few cell surface receptors of bile acids have been discovered so far. TGR5 is a membrane-bound G-protein-coupled receptor for bile acids that are found in several types of tissues types, including liver. TGR5 is involved in a number of important biological processes, such as controlling the energy balance, insulin and glucose homeostasis, inflammation, bile formation and secretion, intestinal motility and secretion, and bile acid -evoked itch and analgesia. The neurons of the enteric and central nervous system also express TGR5 and it is involved in the intestinal motility and detection of endogenous neurosteroids in response to bile acids. The role of TGR5 against metabolic, inflammatory and digestive diseases is becoming prominent and the agonists of TGR5 may promote energy expenditure and insulin release, inflammation, and promote colon transit. Many of the recent reviews have described the conditions in which TGR5 could be a promising new target for pharmaceutical agents. Hence, it becomes important to delineate how these agonists regulate TGR5. The expression and/or activities of P2Ys are highly altered in cells and tissues during the disease progression and more than one P2Ys are involved in many cases. The growing bodies of evidences from experimental and clinical studies emphasize the potential therapeutic value of members of P2Y. As they are considered to be a new therapeutic strategy, several pharmacological agents targeting P2Y receptors are presently available in the market and many of them are under clinical trials. This present chapter summarize the present knowledge on therapeutic significance of TGR5 and P2Y receptors and to reveal the plausible therapeutic use of agents targeting these receptors in the treatment of human diseases.

Keywords Allosteric ligands · Bile acids · Glucose homeostasis · Inflammation · Metabolic diseases

S. Rajagopal, M. Ponnusamy, *Metabotropic GPCRs: TGR5 and P2Y Receptors in Health and Diseases*, https://doi.org/10.1007/978-981-13-1571-8_4

4.1 Introduction

GPCRs are the largest receptor superfamily in the human genome. They are integral part of many important physiological functions. GPCRs have substantial involvement in the pathophysiology of a wide range of human diseases, which includes metabolic, cardiac, immune system and inflammatory disorders. The unique tissue distribution, ligand binding specificity and diversity make GPCRs as ideal therapeutic targets. In addition, the majority of hormones and neurotransmitters (>80%) exert their effects through interacting with GPCRs, which highlights the pharmacological importance of GPCRs as therapeutic targets (Hauser et al. 2017). Thus, GPCRs are most insensitively studied therapeutic targets and this filed rapidly progresses in recent years. A vast number of biological drugs, antagonists, allosteric modulators and biased agonists targeting GPCRs are currently under different phases of drug development. To date, at least one-third of FDA approved drugs can directly or indirectly targeting GPCRs. Based on the list of FDA approved drugs, about 475 of them targets 108 GPCR receptors and 20% of the pharmacological agents currently under clinical trials have the potential to target novel GPCRs (Hauser et al. 2017, 2018). A recent analysis based on approved drugs listed in major databases such as CHEMBL, IUPHAR and DRUGBANG shows that approximately 700 approved drugs directly or indirectly target 134 GPCRs (Fig. 4.1). This number constitutes ~16% of GPCRs in human genome (Sriram and Insel 2018). This list grows every year due to rapid progression of this field of research and new

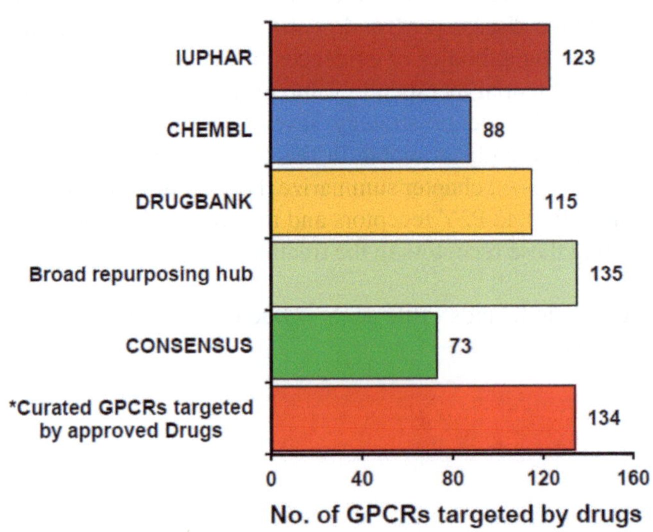

Fig. 4.1 Number of GPCRs targeted by approved drugs listed in major public data base. *Curated analysis was carried out and published by Sriram and Insel (2018)

drugs coming to market. The most commonly used drugs such as α- and β-blockers, β agonists, dopamine receptor blockers, angiotensin receptor blockers, opiates, antihistamines, and selective serotonin reuptake inhibitors can also target GPCRs (Whalen et al. 2011). For instance, the top selling asthma drug (Advair Diskus; fluticasone propionate and salmeterol) and antipsychotic drug (Abilify; Aripiprazole) partially target GPCR. However, a recent genomic study found that some of them could cause adverse effects in the human population depending on the genetic variation of GPCRs in the functional regions, specifically drug- and effector-binding sites (Hauser et al. 2017, 2018).

The current strategy in the drug discovery involves identifying the agents has the capability to enhance or suppress the signal transduction cascade of GPCRs in a particular cell type and tissues. This approach helps to develop drugs to treat GPCR associated disorders. The main target in the traditional GPCR drug discovery is the highly conserved orthosteric sites (the endogenous ligand binding sites), which are located in the 7TM domains of all members of GPCR subtypes. By this approach, many compounds, either directly activate (agonists) or block the endogenous ligand binding by competing with them (antagonists or inverse agonists) are identified. However, the drugs targeting orthosteric sites cause difficulties in screening and achieving high selectivity due to high sequence homology of this region in subtypes (Wootten et al. 2013). The advancement of structural biology and protein engineering have provided the platform for the clear understanding of the structural signature of GPCRs and it revealed that each GPCR subfamily contains structurally distinct allosteric sites, a ligand binding sites distinct from the endogenous ligand binding, within or outside of 7TM domain. The allosteric ligands modulate GPCR activity by stimulating structural conformational changes, which in turn causes changes in orthosteric sites or directly altering effector coupling sites. Depends on their binding nature and activity, they are classified as intrinsic agonist (positive modulator), inverse agonist (negative modulator or neutral allosteric modulator) (Negus 2006; Vauquelin and Van Liefde 2005). Another class of ligand can act on both allosteric and orthosteric sites, which is called as dualsteric or bitopic ligands. Interestingly, bitopic ligands change their affinity and selectivity to allosteric or orthosteric sites depend on the extracellular environment (Lane et al. 2017). The major advantage of allosteric modulators are that they have differential selectivity and high specificity to GPCR subtypes due to high variance in the amino acid sequence of this region, which enables the fine tuning of cellular responses without affecting the temporal and spatial features of physiological effects mediated by endogenous ligands. The recent X-crystallography studies of class A, B, C and F have identified that several GPCRs including P2Y1 possess sites for allosteric modulators (Wootten et al. 2013). Thus, the diversity and a distinct topography of allosteric sites have offered a new dimension of GPCR signalling manipulation and enriched the way of screening and development of highly potent and selective therapeutic agents targeting a particular GPCR (Fig. 4.2).

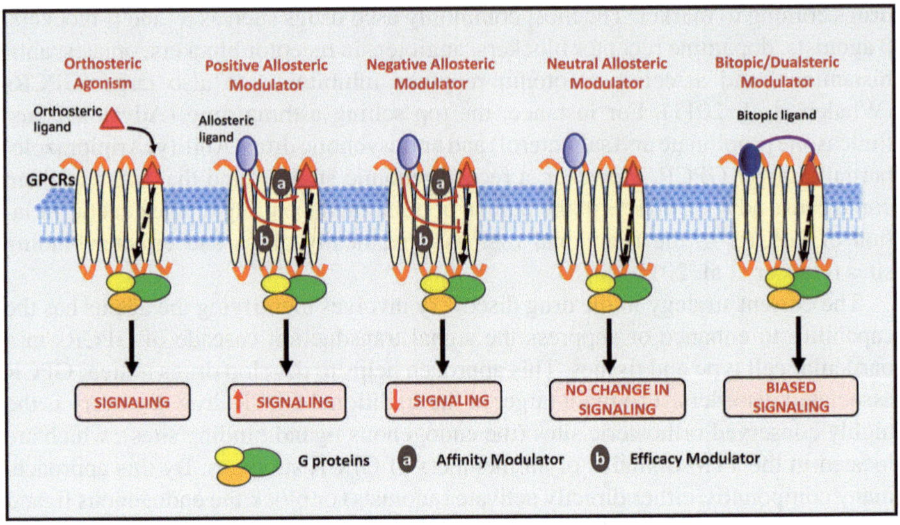

Fig. 4.2 Mode of action of allosteric ligands and their influence on GPCR signaling. The binding of orthosteric agonists induces conformational changes in GPCR and results in the activation of downstream signaling. The allosteric modulators bind to a site distinct from orthosteric sites and influence GPCR downsteam signaling. The binding of positive allosteric modulators enhances the affinity (cooperativity factor-a) and/or efficacy (modulation factor-b) of the orthosteric agonist, while the negative allosteric modulators reduce the affinity (cooperativity factor-a) and/or efficacy (modulation factor-b) of the orthosteric agonist. The neutral allosteric ligands do not affect the affinity and/or efficacy of orthosteric agonists. Bitopic or dualsteric modulators are bilvalent ligands, which bind to both allosteric and orthosteric sites. These ligands composed of orthosteric moiety (agonist or antagonist) linked to allosteric moiety (positive or negative modulator) through a chemical linker. The arrow heads represent activation. The blunded arrows represent inhibition

4.2 Therapeutic Potential of TGR5

TGR5 expression has been found in humans and animals and it is recognized as the first known G-protein coupled receptor specific for bile acids. The differential experiments have revealed that the activation of TGR5 regulates the basal metabolism and energy expenditure. This receptor is a potential target in the treatment for metabolic disorders, such as type 2 diabetes (Hodge and Nunez 2016; Michal et al. 2018). TGR5 has an ability to lower the proinflammatory cytokines and it opens a new vision in the pathophysiology of inflammatory processes associated disorders such as colitis and atheroma development. The expression of TGR5 gene decreased after incubation for 72 hr. with potent TGR5 ligand steroid 5-beta-pregnan-3-alpha-ol-20-one (Michal et al. 2018; Kwang-Hoon et al. 2009; Pols et al. 2014).

Cholangiocytes secrete bicarbonate and it is responsible for the gallstone formation from the cholesterol which are stimulated by TGR5 activation. The bile acids dependent activation of TGR5 has influence on energy balance, inflammation, digestion, and sensation. cAMP formation is stimulated by the primary and secondary bile acids such as DCA, taurolithocholic acid (TCA) and the selective agonists

OA and 3-(2-chlorophenyl)-N-(4-chlorophenyl)-N, 5-dimethylisoxazole-4-carboxamide (CCDC). The understanding of the spatiotemporal control of TGR5 signaling provides insights into the actions of bile acids and therapeutic potential of TGR5 agonists/antagonists. The cholecystokinin is a hormone known to induce the contraction of gall bladder and secretion of bile into the intestine that is released by the stimulation of lipids present in the small intestine from the food (Handelsman 2011; Thomas et al. 2009). Bile contents are primarily composed of CDC and CA, which are the two primary bile acids in humans that aid the emulsification and absorption of dietary fats and the lipid-soluble vitamins A, D, E, and K. However, if there are any defects in bile acid metabolism, which can cause metabolic disorders in human. Neurosteroids are potential TGR5 agonists present in the central nervous system that are structurally related to bile acids (Yanguas-Casás et al. 2017; Kim et al. 2017). Although the signal transduction mechanisms of TGR5 are currently not completely understood, the broad range of biological actions of TGR5 has provided an impetus for the development of agonists and antagonists targeting TGR5 for the treatments of metabolic, inflammatory, digestive, and sensory disorders (Lieu et al. 2014; Pols et al. 2014).

4.2.1 TGR5 Agonists

Taurine, glycine-conjugated bile acids and unconjugated bile acids belong to the steroids family and that are the most powerful endogenous agonists of TGR5 receptor. TGR5 has differential affinity for various bile acids, with the following rank order of potency: lithocholic acid ≥ deoxycholic acid > chenodeoxycholic acid > cholic acid. Taurine-conjugated bile acids strongly activate TGR5 compared to unconjugated bile acids and the response from this activation is more effective than glycine-conjugated bile acids. It is well known that the conjugation of taurine with bile acids is the default process in non-mammalian orders. While the glycine conjugation is uncommon in many mammals and it occurs mostly in bovids and primates (Gertzen et al. 2015).

Many natural compounds mimic bile acids and they can stimulate TGR5. Oleanolic acid (OA) is a triterpene molecule from olive leaves, which resemebles the structure of lithocholic acid and it could act as an agonist of bile acid receptor (Wang et al. 2017). The recent studies have shown that TGR5 agonists derived from natural C_{24} bile acids to be a more potent selective agonist of TGR5, in addition to bile acids and OA (Sato et al. 2008). So far, S-EMCA (6α-ethyl-23(S)-methylcholiacid) is the only most specific agonist available, which is synthesized from cholic acid. Similarly, INT 767, a modified bile acid derivative, which is known to activate both TGR5 and FXR (Pathak et al. 2017). Further, a new therapeutic targets for type 2 diabetes, 3-aryl-4-isoxazolecarboxamides have been developed from other synthetic agonists and they are non-bile-acid-related compounds, but they have high affinity to human TGR5 The class of the 2-aryl-3-aminomethylquinolines have also affinity for TGR5 (Evans et al. 2009). However, these agonists have not been tested in clinical settings (Figs. 4.3 and 4.4).

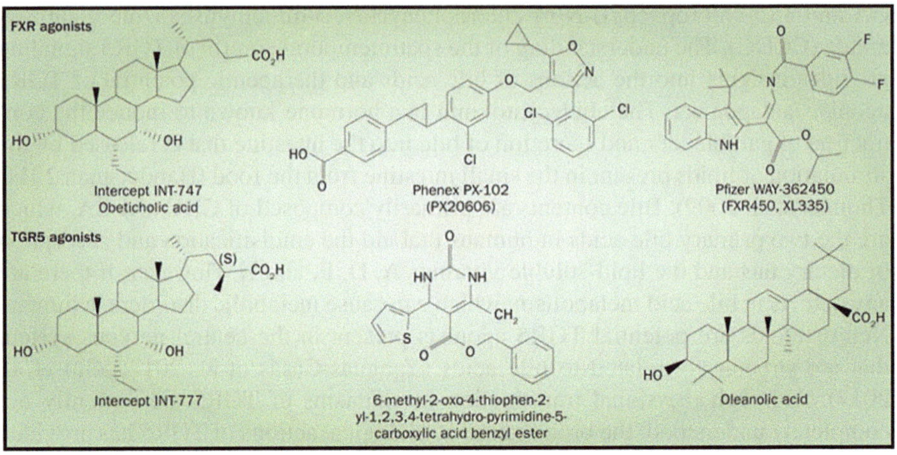

Fig. 4.3 Agonists for plasma membrane receptor TGR5, and Nuclear membrane receptor FXR

4.2.2 Functional Effects of TGR5 Agonists

The circulatory levels of Bile acids may be decreased or increased during feeding and fasting, and during cholestatic diseases, and the cells would be continuously exposed to high levels of bile acids. Many GPCRs signaling activities are desensitized and they are trafficked to endosomes immediately after activation. However, the activated TGR5 is retained in plasma membrane lipid rafts, from where it can transmit sustained signals. Currently, the molecular mechanisms controlling the short-term signaling of TGR5 largely unknown, and the impact of long-term exposure of TGR5 to agonists at the transcriptional level remains to be determined.

CCDC is a synthetic agonist for TGR5 that can stimulate the association of TGR5 with β-arrestin2 and GRK2, and this is opposite to the actions of endogenous bile acids. Despite CCDC is a biased agonist of TGR5, this agonist can stimulate β-arrestin2 and GRK2 recruitment, which indicates its increased potency. Apart from this, it is also important to know whether TGR5 agonists have differential effect on TGR5-mediated signaling, receptor trafficking and their functions as biased agonists. A recent study on the role of TGR5 in cholestatic disease associated pruritus indicate that the antagonists of this receptor could provide a relief from itch (Dawson 2011).

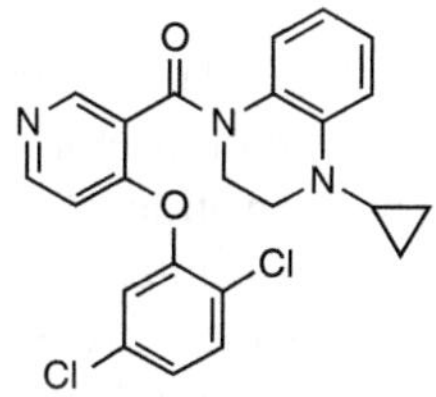

1a, 2 nmol/L (hTGR5)

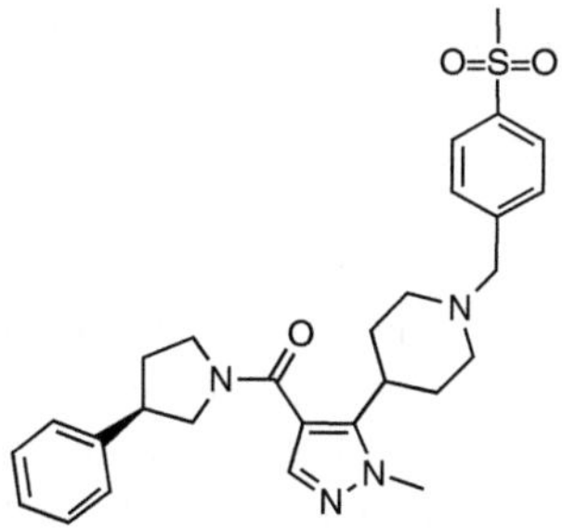

1b, 3 nmol/L (hTGR5)

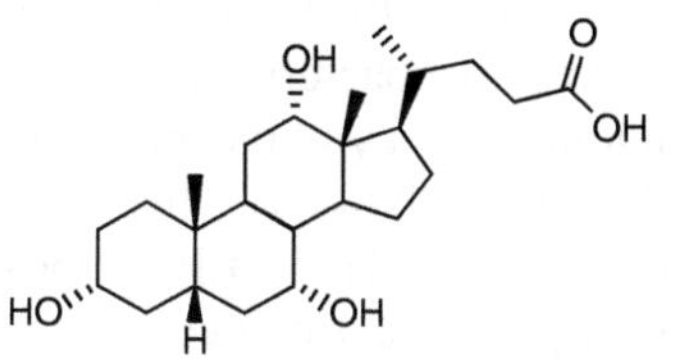

2a, Cholic acid,
13.6 μmol/L (hTGR5)

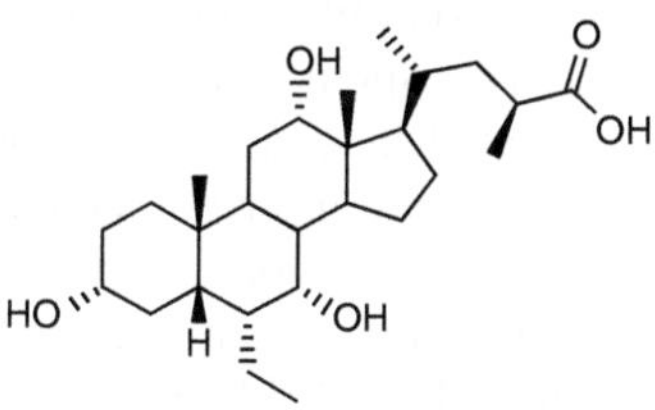

2b, INT777,
820 nmol/L (hTGR5)

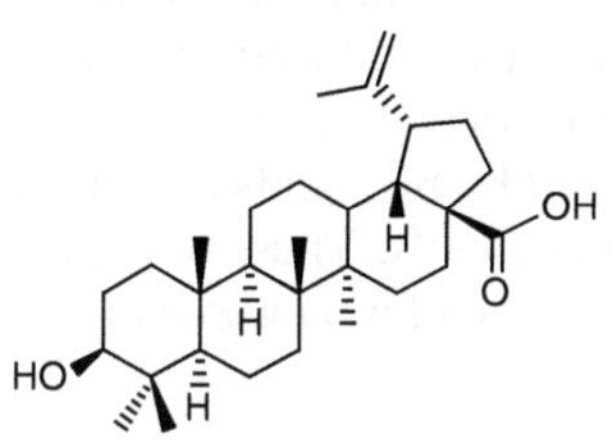

3β, Betulinic acid,
2.25 μmol/L (hTGR5)

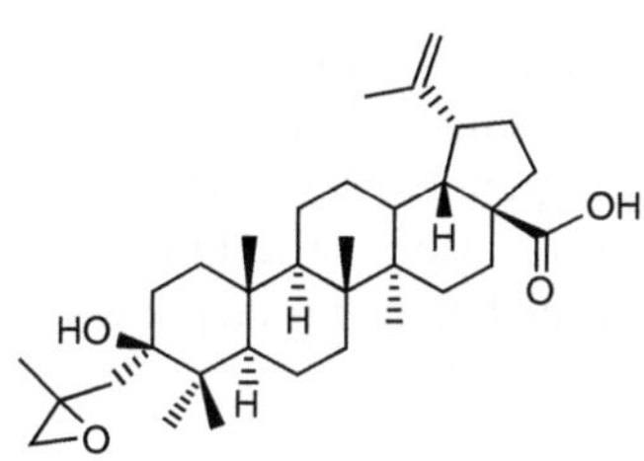

4, 47 nmol/L (hTGR5)

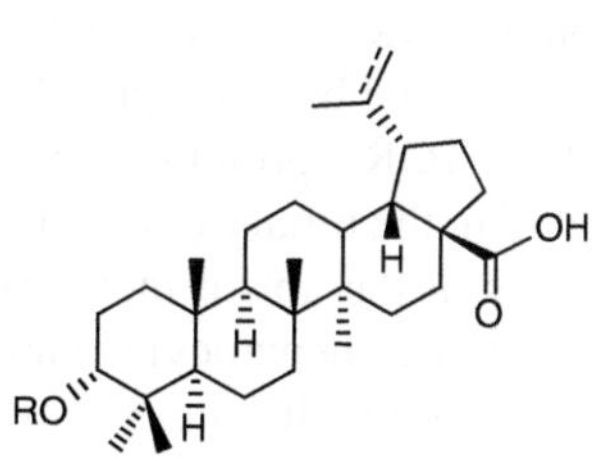

R=H or Acyl group

Fig. 4.4 Known TGR5 agonist and their EC value

4.2.3 Biological Effects of TGR5 Activation

4.2.3.1 TGR5 Activation on Inflammation

TGR5 is expressed in human monocytes (Duboc et al. 2014). Several studies suggest that TGR5 is involved in the previously known immunosuppressive effect of bile acids in immune cells (Perino et al. 2014). The activation of TGR5 leads to suppression of levels of pro-inflammatory cytokines such as interleukin-1α (IL-1α), IL-1β, IL-6 and tumour necrosis factor-α (TNF-α), which are synthesized by Kupffer cells in response to a bacterial endotoxin such as lipopolysaccharide (LPS) (Perino et al. 2014). This is confirmed in mice with deficiency of TGR5 expression in which LPS stimulates profound increase of the pro-inflammatory cytokines. These findings reveal that TGR5 signalling is required for the down regulation of the inflammatory response triggered by Gram-negative bacteria (Kaia et al. 2014).

TGR5 mediated blockage of the pro-inflammatory cytokine production occur through inhibition of transcriptional nuclear factor-κB (NF-κB). This is an evidence for TGR5 anti-inflammatory property (Kaia et al. 2014). In LPS exposed cells, TGR5 dependent suppression of NF-κB transcriptional activity can be reversed by cAMP inhibitors (Pols 2014). In addition, TGR5 activation by bile acids can influence the differentiation of subtypes of dendritic cells into cells with an anti-inflammatory phenotype (Ichikawa et al. 2012). Due to their role as antigen presenting cells, dendritic cells are critical regulators of the acquired and innate immunity. In experimental studies, the exposure to bile acid causes reduced production of IL-12 during the differentiation of dentritic cells from monocytes when it is stimulated by bacterial antigen. However, these effects induced by bile acids or a specific TGR5-agonist do not occur in terminally differentiated cells. Thus, the upregulation of TGR5 is crucial in the initial phase of differentiation of monocytes to dendritic cells (Ichikawa et al. 2012).

4.2.3.2 TGR5 Activation in Liver Physiology

The endothelial cell expression of TGR5 activates nitric oxide synthase (NOS) and induces NO release. It is also expressed by Kupffer cells and liver sinusoidal endothelial cells (Keitel et al. 2007). TGR5 protein is not detectable in hepatocytes (Keitel et al. 2013), but it is found in the biliary tree and gallbladder epithelia (Keitel and Häussinger 2011), as well as in the apical membranes and primary cilium of cholangiocytes (Pols et al. 2014). The hepatocytes can synthesis and secrete bile acid but it does not express TGR5, while the site of action of bile acid is the biliary tree, where TGR5 is abundant, which represents paracrine action of bile acids within this organ (Pols et al. 2011). In the human gallbladder cells, TGR5 present in cystic fibrosis transmembrane conductance regulator (CFTR) and the apical sodium dependant bile salt uptake transporter (ASBT) (Bijvelds et al. 2005). An experimental study using CFTR inhibitor and TGR5 agonist found that TGR5 activates CFTR and promotes biliary chloride secretion (Bijvelds et al. 2005). In addition, TGR5 should be a key player in this sensing mechanisms of the primary cilium, which acts a sensor to regulate the bile flow and composition. The genetic deletion of TGR5 in mice confirmed this notion. In those mice, the TGR5 knock out declines the usual increase of bile flow, chloride, and bicarbonates secretion after partial hepatectomy in mice and it also results in a slow regeneration of the liver (Duboc et al. 2014). In cholangiocytes, TGR5 is distributed in different subcellular compartments and it could have differential effects depending on its complex cellular repartition (Masyuk et al. 2013). A functional study in mice showed that TGR5 stimulates gallbladder filling along with reduction of motility (Li et al. 2011). An *ex vivo* experiments confirmed that TGR5 activation leads to reduction of contractile properties of smooth muscles in the gallbladder. Thus, TGR5 has an important role in stimulating gallbladder smooth muscle relaxation and filling (Li et al. 2011).

4.2.3.3 Regulation of TGR5 in Energy Homeostasis and Metabolism

TGR5 enhances energy expenditure, increases oxygen uptake, blocks obesity, and reduces insulin resistance in animal model of obesity (Watanabe et al. 2009). TGR5 promotes the conversion of thyroxin (T4) into intracellular active tri-iodothyronine (T3), a major component involved in cellular basal metabolism, by activating the key enzyme iodothyronine deiodinase in human brown adipocytes and skeletal myocytes (Watanabe et al. 2009). In addition, TGR5 protects against insulin resistance by triggering the intestinal secretion of glucagon-like peptide 1 (GLP-1), which is evident fromTGR5 agonists induced increase of GLP-1 release from a enteroendocrine cells line (Bala et al. 2014). Moreover, TGR5 is expressed by pancreatic cells islets of mammals including mouse and human pancreatic cells islets and its stimulation with agonists can induce cAMP/Ca^{2+} pathway and increases insulin secretion (Kumar et al. 2012), which confirms the influence of bile acids in glucose homeostasis. The natural agonist of TGR5 receptor, OA, enhances glucose tolerance by increasing insulin secretion and reduction of blood glucose level in a mouse model of obesity and diabetes (Chen et al. 2011). Together, these studies indicate that bile acid–TGR5 acts as a key regulator of basal metabolism regulation (Guo et al. 2016).

4.2.3.4 TGR5 Regulation on Intestinal Motility and Secretion

TGR5 contributes to the gastrointestinal motility (Bunnett and Cottrell 2010). The genetic ablation of TGR5 causes impairment in peristalsis movement, delayed colonic transit and a lower defecation frequency. A mechanistic study revealed that bile acids promote the release of the peristaltic transmitters such as 5-hydroxytryptamine and calcitonin-gene related peptide. In fact, the deletion of TGR5 in intestine does not affect mechanically induced peristalsis, which clearly indicates that colon transit is specifically induced by bile acids. These findings suggest that TGR5 plays a physiological role in the regulation of propulsive motor function (Alemi et al. 2013).

In addition, the water content of faeces is reduced in TGR5 knockout mice, which suggests that TGR5 is critically involved in the regulation of the colonic water and electrolytes efflux. However, it would not possible to define its role in chloride secretion in intestine without a functional assay using a specific TGR5 inhibitor.

4.2.3.5 Association of TGR Activation in Gastrointestinal Cancer

Several epidemiological studies suggest that there is a link between high levels of faecal bile acids and colon cancer (Hana et al. 2014), since bile acids has possible carcinogenic role in colon cancer (Hill 1990). However, the debate still remains due to the dual effects of bile acids on apoptosis and cell proliferation. Nonetheless, several studies show that the excessive bile acids in the intestine, as a dietary supplementation of deoxycholic acid, (at a level of 0.2%), induces colon cancer in mice. A molecular study found that bile acid-induced suppression of miR-199, which targets CAC1, a cell cycle-related protein, and thereby TGR5 acts as a tumour suppressor (Yasuto et al. 2014; Dufer et al. 2012).

In general, it is believed that TGR5 is not associated with colonic carcinogenesis. However, two *in vitro* studies have shown its implication in gastric and oesophageal carcinoma (Phelan et al. 2017; Yui et al. 2008). In humans, distal gastrectomy causes several complications including gastroduodenal reflux. As the intestinal contents have bile acids, the gastroduodenal reflux of bile acids into the stomach and oesophagus increases the risk of oesophageal and gastric cancer (Jenkins et al. 2008). It is proved in gastric carcinoma cell lines in which bile contents activate carcinogenic pathways (Shreiner et al. 2015). Moreover, the upregulation of TGR5 activity stimulates proliferation of Barrett's oesophagus, which is an oesophageal metaplasic condition well-known for its risk of developing into oesophageal adenocarcinoma (Shreiner et al. 2015). A recent clinical study found that TGR5 increases in patients with gastric adenocarcinoma and increased TGR5 level significantly related to the worse patient survival. In contrast, a report from Chen et al. demonstrate that TGR5 should have a protective role in human hepatocellular carcinoma (Min-Chan et al. 2016).

4.2.3.6 Intracellular Signalling Cascades Induced by TRG5 Activation in Eukaryotic Cells

The bile acid-dependent activation of TGR5 primarily results in the activation of the adenylyl cyclase cAMP signalling pathways (Rajagopal et al. 2015), but it also activates other types of cell signalling cascades depends on the cell type. For example, the ligands of TGR5 such as OA and S-EMCA selectively activates Gα(s) in the pancreatic β-cell line (Faria et al. 2009), which increases cellular level of cAMP, calcium, and phosphoinositide hydrolysis, and this can be blocked by NF449 (a selective Gα (s) inhibitor) or U 73122 (PI hydrolysis inhibitor) (Kumar et al. 2012). Similarly, cAMP level increases along with a rise in intracellular calcium in cultured neurons that could stimulate the production of reactive nitrogen oxide species (Bala et al. 2014). The activation of TGR5 in liver sinusoidal endothelial cells increases cAMP release and it subsequently activates eNOS by phosphorylation, which is responsible for the nitric oxide (NO) production (Kida et al. 2013). In contrast to its effect in neurons and β-cells lines, TGR5 lowers the intracellular calcium levels in gallbladder smooth muscles. It is well known that decreased rhythmic discharge of intracellular Ca^{2+} is necessary to induce contraction (Lavoie et al. 2010; Wang et al. 2011) .

4.3 P2Y Receptors and Their Therapeutic Potential

The ubiquitous distribution of P2Y receptors in many organs and their contribution in many physiological functions have implicated them in the pathophysiology of a wide range of diseases including central nerves system, cardiovascular system, kidney, liver, pulmonary system, digestive system and metabolic diseases. The expression and/or activities of P2Ys are highly altered in cells and tissues during the disease progression and more than one P2Ys are involved in most cases. The growing bodies of evidences from experimental and clinical studies emphasize the potential therapeutic value of members of P2Y. As they are considered as a new novel therapeutic targets, several pharmacological agents targeting P2Y receptors are already available in the market and some of them are under clinical trials (Burnstock 2016; Burnstock and Loesch 2017). One classical example is clopidogrel, a P2Y12 antagonist, is used for the treatment of thrombosis and stroke, despite some patients have response variability (Cattaneo et al. 2011). Diquafosol (a uridine nucleotide analogue) is a P2Y2 receptor agonist, which is used in clinical settings for the management and treatment of dry eye disease (Lau et al. 2014). The following table shows list of some commonly used drugs targeting P2Y receptors and their therapeutic use in clinical conditions (Tables 4.1 and 4.2).

Table 4.1 Most commonly used drugs and their targets

Drug	Target	Mode of action	Clinical use
Prasugrel, Cangrelor, Elignogrel, Clopidogrel	P2Y12	Antagonist	Antithrombosis, stroke and cardiovascular disease
Risperidone, clozapine	P2Y1, P2Y12	Antagonist	Antipsychotic drugs, Antithrombosis
Haloperidol	P2Y1	Antagonist	Antipsychotic drug, Antithrombosis
Diquafosol	P2Y2	Agonist	Dry eye disease, cystic fibrosis

Table 4.2 List of agonist and antagonist targeting P2Y receptors

Chemical/ pharmacological agent name	Target	Therapeutic use	References
Antagonists			
MRS 2179, (N)-methanocarba-2MeSAD	P2Y1	Antiplatelet aggregation, thrombosis	Gremmel et al. (2016) and Chhatriwala et al. (2004)
MRS2500	P2Y1	Neuroprotective	do Carmo et al. (2014)
GLS409	P2Y1 and P2Y12	Antiplatelet aggregation, thrombosis	Gremmel et al. (2016)
SAR216471	P2Y12	Antiplatelet aggregation, thrombosis	Delesque-Touchard et al. (2014)
AR-C118925	P2Y2	Anticancer, Antiinflammation, renal disease	Jiang and Dingledine (2013)
MRS2578	P2Y6	Antifibrosis, Antineurtophil activation, Antiinflammation	Muller et al. (2004) and Sil et al. (2017)
NF340	P2Y11	Antitumor, Antifibrosis	Khalid et al. (2017) and Dreisig and Rahbek (2016)
Agonists			
MRS2365	P2Y1	Antiplatelet aggregation, Blood pressure regulation	Bourdon et al. (2006) and Wenker et al. (2013)
MRS2768	P2Y2	Cardioprotective	Hochhauser et al. (2013)
ATP-γ-S	P2Y2	Refractory gastroesophageal reflux disease	Wu et al. (2017)
Denufosol tetrasodium (INS37217)	P2Y2	Cystic fibrosis in lungs	Burch and Picher (2006)

4.3.1 P2Y as a Target of Cardiovascular Diseases (CVD)

Emerging evidences indicate that P2Y receptors can functions in a non-classical way in cardiovascular system, without its nucleotide ligand binding, through dimerization with GPCRs. This phenomenon has made P2Y receptors as a new attractive therapeutic target for treating CVD. Many experimental studies have proved the significance of targeting P2Y receptors using various pharmacological agents and genetically modified experimental animal models. The vascular endothelial expression of P2Y1 and P2Y2, and their downstream effector G-proteins (Gq/G11) control the blood pressure by promoting vasodilation through stimulation of nitric oxide release, which results in a decrease of blood pressure. Normally, ATP and UTP are released from endothelial cells during shear stress caused by changes in blood flow that induce vascular relaxation by acting in a paracrine manner (Burnstock and Loesch 2017; Wang et al. 2015). The experimental studies using agonists of P2Y1 and P2Y2 have shown reduced blood pressure and improved circulation. The systemic administration of INS45973 (Ip4U·4 Na+), a UTP analog and activator of P2Y2, effectively reduced acute blood pressure in mice and this observation was confirmed using P2Y2 knockout mice (Rieg et al. 2011). An experimental study found that pretreatment with MS2678 (P2Y2 agonist) protects cardiomyocytes from hypoxia and prevents cardiac damage caused by myocardial ischemia (Hochhauser et al. 2013). The second generation P2Y1 receptor agonists generated from adenosine 5′-O-(1-boranotriphosphate) derivatives (Farret et al. 2006) are also useful to prevent elevation of blood pressure, but it still needs further experimental and clinical evaluation. Likewise, the agonist of P2Y1 (MRS2365) increases the response of sympathetic nerve activity and improves blood pressure in animal model (Wenker et al. 2013).

The UDP/UTP dependent receptor, P2Y6 is involved in many cardiovascular problems. The myogenic tone (MT) of resistance arteries is vital for autoregulation of blood flow under normal circumstances and that relies on contractile properties of vascular smooth muscle cells. However, the enhanced MT is known to cause peripheral vascular resistance, which results in pressure overload, hypertension and coronary heart failure. A study with P2Y6 knockout mice found that the inactivation or absence of P2Y6 protects against pathological increase of MI, blood pressure and CHF caused by angiotensin II or myocardial infarction. P2Y6 dependent activation of $G_{q\text{-}11}$ and RhoA-GTP binding is required for this process. The inhibitors of P2Y6 such as blue-2 and MRS2578 can effectively block these pathological alterations (Kauffenstein et al. 2016). The increased activity of P2Y6- Gα12/13 in cardiomyocytes of heart with pressure overload triggers fibrotic response and cardiac remodeling by upregulating AngII and TGFβ mediated expression of fibrogenic genes. P2Y6 antagonist (MRS2578) has the capability to block the fibrotic response and improve cardiac function (Nishida et al. 2008). In addition, MRS2578 blocks P2Y6-NFKβ dependent vascular inflammation and atherosclerosis (Riegel et al. 2011). The platelets aggregation before percutaneous coronary intervention (PCI; also known as angioplasty with stent) and after cardiac surgery are the common problem in patient with CVD. This aggregation leads to coronary atherosclerosis, ischemic events and MI after surgery due to stent- induced rupturing, bleeding and thrombus. To prevent this, a thienopyridine based drug, clopidogrel, is most commonly used in the therapeutic regimen. However, some patients are not responding to the treatment and they are more prone to ischemic events due to genetic polymorphisms and/or loss of functional CYP2C19, which is required for its metabolism and activation in biological system. The new generation P2Y12 inhibitors such as ticagrelor, cangrelor, elignogrel and prasugrel overcome this problem as they do not require CYP2C19 metabolism (Damman et al. 2012; Zetterberg and Svensson 2016). In addition, a meta-analysis reveals that prasugrel and ticagrelor are well effective in patients with ST-segment elevation myocardial infarction (STEMI) undergoing primary PCI, while prasugrel is better than ticagrelor in conjunction with drug-eluting stents (Cattaneo et al. 2014). The synergetic inhibition of P2Y1 and P2Y12 with GLS409 rapidly and more effectively blocks platelet mediated thrombosis irrespective of asprin intake (Gremmel et al. 2016). Currently, chemical agents are under development in the aspect of high selectivity and more efficacies to target P2Y and treat cardiovascular problems (Fig. 4.5).

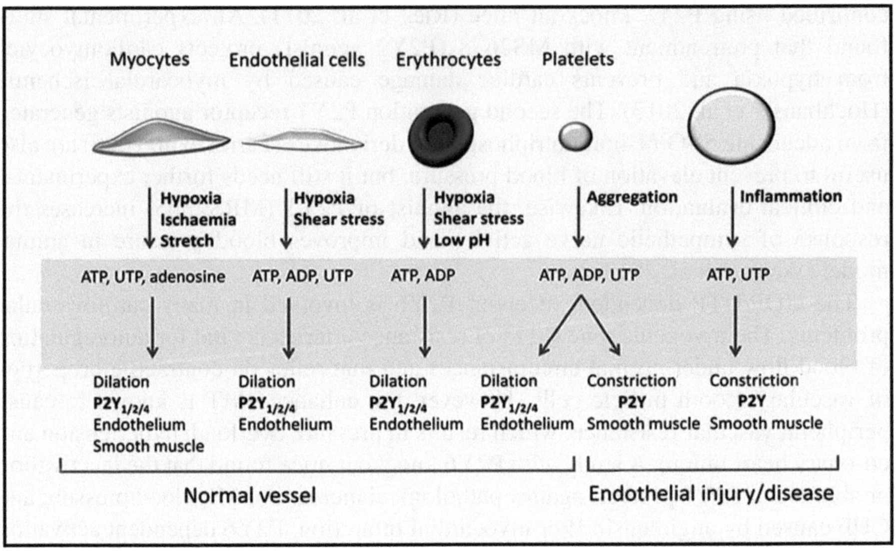

Fig. 4.5 P2Y receptors functions and their implication as drug targets in CVD

4.3.2 P2Y as a Target of Neurological Disorders

The extracellular nucleotides, mainly ATP, are involved in the physiology of neurotransmission and neuromodulation under normal circumstances. However, a large amount of ATP and UTP are released into extracellular environment under pathological condition and endure the abnormal expression and activity of various P2 receptors including P2Y subtypes. It is well documented that P2Y receptors associated with a variety of neurodegenerative diseases including multiple sclerosis, Alzheimer's disease, Parkinson's disease, stroke, migraine, epilepsy, psychiatric disorders and schizophrenia (Burnstock 2016). Epilepsy is a heterogeneous group of neurological disorders, which are characterized by recurrent seizers. The most conventional drugs currently for epilepsy treatment target Ion channels, GABA or glutamate receptors. However, nearly one-third of patients undergoing treatment are having pharmaco-resistance to these classical targets. This has lead to find new targets for epilepsy and P2 receptors are gaining much attention in the epilepsy treatment since the P2 receptors are linked to the neuroinflammatory process, which is a key player in seizure generation. The experimental studies in mice models suggest that a high ATP level in brain leads to high spiking and it provokes seizure severity during epilepticus. Surprisingly, treatment with UTP and its metabolites remarkably suppress epileptic seizure under the condition of epileptic enchepalopathy. As UTP and UDP mainly target P2Y2, P2Y4, P2Y6 and P2Y14, the synthetic agonists of these P2Y receptors could prevent the progression of seizure. A molecular study in rat primary cortical neuron confirmed the neuroprotective response of P2Y2 receptor under upregulated status of proinflammatory processes (Peterson et al. 2010). In addition, P2Y6 play integral role in the migration, chemotactic function and phagocytotic activity of microglia, which indicates the important regulatory role of P2Y6 in neuroinflammation (Ricardo et al. 2015). Thus, the manipulation of receptor activity using exogenous agonist or antagonist can prevent and halt the progression of neurological disorders.

4.3.3 P2Y as a Target of Pulmonary Disorders

The fundamental role of P2Y signaling in the inflammatory process contributes to various disorders in lungs including bronchial asthma and chronic obstructive pulmonary disease (COPD). In both human and animal models, P2Y2 and P2Y6 contribute to the pathogenesis of idiopathic pulmonary fibrosis (IPF). A strong upregulation of P2Y2 and P2Y6 in macrophages and neutrophils of IPF patients elicits proinflammatory response, fibroblast activation and pulmonary fibrosis (Albalawi et al. 2017). The inhibition of P2Y6 with MS2578 blocks inflammatory and pro-fibrotic processes (Albalawi et al. 2017). On the other hand, the increased activity of P2Y2 receptor can alleviate cystic fibrosis, which is caused by a defect in chloride channel (CFTR). The activators of P2Y2 such as denufosol increases clearance of mucociliary and improve lung function by activating alternative chloride channels other than CFTR (Schwiebert et al. 2005). A new generation P2Y2 agonist, Denufosol tetra sodium (INS37217), has more advantages in terms of stability and long half life than other P2Y2 activators, which enables the extended duration of its action and substantial clearance of mucin release and clearance (Faria et al. 2009).

4.4 Future Perspective

TGR5 could be a promising target to enhance several aspects of the metabolic syndrome. Experimental studies revealed that TGR5 has many beneficial effects including reduction of body weight improvement of glucose homeostasis and reduces hepatic steatosis. The anti-inflammatory action of TGR5 on macrophage-driven inflammation should also considered as a potential positive effect of TGR5 with regard to the metabolic syndrome. These properties of TGR5 clearly suggest that activation of this GPCR is valuable within the framework of the metabolic syndrome. Similarly, the purine- and pyrimidine-sensitive P2Y receptors are linked to broad range of physiolocal and pathological functions including blood clotting, cardiac function, refractory gastroesophageal reflux disease, inflammation and tumorigenesis. Several P2Y based drugs such as Clopidogrel, Clozapine, Diquafosol, etc. are still used as standardised therapeutic regimen to treat patients. Therefore, it is not unexpected that the P2Y receptors could be useful targets for drug development. As the growing evidence showing the important of TGR5 and P2Y receptors in the health of humans for many diseases, we can expect more beneficial and efficient drugs targeting P2Y as well as TGR5 receptors in the future. The identification and development of several natural, semisynthetic and synthetic antagonist or agonists would further advance these receptor as a valid target for treatment. But, still much work needs to be carried out to clarify the safety, specificity and efficacy.

References

Albalawi, F., Lu, W., Beckel, J., Lim, J., McCaughey, S., & Mitchell, C. (2017). The P2X7 receptor primes IL-1β and the NLRP3 inflammasome in astrocytes exposed to mechanical strain. *Frontiers in Cellular Neuroscience, 11*, 227.

Alemi, F., Poole, D., Chiu, J., Schoonjans, K., Cattaruzza, F., Grider, J., Bunnett, N., & Corvera, C. (2013). The receptor TGR5 mediates the Prokinetic actions of intestinal bile acids and is required for normal defecation in mice. *Gastroenterology, 144*, 145–154.

Bala, V., Rajagopal, S., Kumar, D. P., Nalli, A. D., Mahavadi, S., Sanyal, A. J., Grider, J. R., & Murthy, K. S. (2014). Release of GLP-1 and PYY in response to the activation of G-protein coupled bile acid receptor TGR5 is mediated by Epac/PLC e pathway and modulated by endogenous H_2S. *Frontiers in Physiology, 5*, 1–11.

Bijvelds, M., Jorna, H., Verkade, H., Bot, A., Hofmann, F., Agellon, L., Sinaasappel, M., & de Jonge, H. (2005). Activation of CFTR by ASBT-mediated bile salt absorption. *American Journal of Physiology; Gastrointestinal and Liver physiology, 289*, G870–G879.

Bunnett, N., & Cottrell, G. (2010). Trafficking and signaling of G protein-coupled receptors in the nervous system: Implications for disease and therapy. *CNS Neurological Disorders and Drug Targets, 9*, 539–556.

Burch, L., & Picher, M. (2006). E-NTPDases in human airways: Regulation and relevance for chronic lung diseases. *Purinergic Signalling, 2*, 399–408.

Burnstock, G. (2016). Purinergic signalling in the gut. *Advanced Experimental and Medical Biology, 891*, 91–112.

Burnstock, G., & Loesch, A. (2017). Sympathetic innervation of the kidney in health and disease: Emphasis on the role of purinergic cotransmission. *Auton Neuroscience, 204*, 4–16.

Bourdon, D. M., Mahanty, S. K., Jacobson, K. A., Boyer, J. L., & Harden, T. K. (2006). (N)-methanocarba-2MeSADP (MRS2365) is a subtype-specific agonist that induces rapid desensitization of the P2Y1 receptor of human platelets. *Journal of Thrombosis and Haemostasis, 4*, 861–868.

Cattaneo, G., Podda, G., & Cattaneo, M. (2011). Recent advances on the studies of the platelet's inhibition and aggregation. State of the art of new P2Y12 antagonists. *Recent Progressive Medicine, 102*, 150–155.

Cattaneo, F., Guerra, G., Parisi, M., De Marinis, M., Tafuri, D., Cinelli, M., & Ammendola, R. (2014). Cell-surface receptors transactivation mediated by g protein-coupled receptors. *International Journal of Molecular Sciences, 15*, 19700–19728.

Chen, X., Lou, G., Meng, Z., & Huang, W. (2011). TGR5: A novel target for weight maintenance and glucose metabolism. *Experimental Diabetes Research, 2011*, 1–5.

Chhatriwala, M., Gnana Ravi, G., Patel, R., Boyer, J., Jacobson, K., & Kendall Harden, T. (2004). Induction of novel agonist selectivity for the ADP-activated P2Y1 receptor versus the ADP-activated P2Y12 and P2Y13 receptors by conformational constraint of an ADP analog. *Journal of Pharmacology and Experimental Therapeutics, 311*, 1038–1043.

Damman, P., Woudstra, P., Kuijt, W., de Winter, R., & James, S. (2012). P2Y12 platelet inhibition in clinical practice. *Journal of Thrombosis and Thrombolysis, 33*, 143–153.

Dawson, P. (2011). Role of the intestinal bile acid transporters in bile acid and drug disposition. *Handbook of Experimental Pharmacology, 201*, 169–203.

Delesque-Touchard, N., Pendaries, C., Volle-Challier, C., Millet, L., Salel, V., Hervé, C., Pflieger, A., Berthou-Soulie, L., Prades, C., Sorg, T., Jean-Marc, H., Savi, P., & Bono, F. (2014). Regulator of G-protein signaling 18 controls both platelet generation and function. *PLOS One, 9*(11), e113215.

do Carmo, J., da Silva, A., Ebaady, S., Sessums, P., Abraham, R., Elmquist, J., Lowell, B., & Hall, J. (2014). Shp2 signaling in POMC neurons is important for leptin's actions on blood pressure, energy balance, and glucose regulation. *American Journal of Physiology: Regulatory and Integrative Compartive Physiology, 307*, R1438–R1447.

Dreisig, K., & Rahbek, B. (2016). A critical look at the function of the P2Y11 receptor. *Purinergic Signalling, 12*, 427–437.

Duboc, H., Taché, Y., & Hofmann, A. (2014). The bile acid TGR5 membrane receptor: From basic research to clinical application. *Digestive Liver Diseases, 46*, 302–312.

Dufer, M., Hörth, K., Krippeit-Drews, P., & Drews, G. (2012). The significance of the nuclear farnesoid X receptor (FXR) in beta cell function. *Islets, 4*, 333–338.

Evans, K., Budzik, B., Ross, S., Wisnoski, D., Jin, J., Rivero, R., Vimal, M., Szewczyk, G., Jayawickreme, C., Monco, D., Rimele, T., Armour, S., Weaver, S., Griffin, R., Tadepalli, S., Jeune, M., Shearer, T., Chen, Z., Chen, L., Anderson, D., Becherer, J., De Los Frailes, M., & Javier Colilla, F. (2009). Discovery of 3-Aryl-4-isoxazolecarboxamides as TGR5 receptor agonists. *Journal of Medicinal Chemistry, 52*, 7962–7965.

Faria, D., Schreiber, R., & Karl, K. (2009). CFTR is activated through stimulation of purinergic P2Y2 receptors. *Pflügers Archiv – European Journal of Physiology, 457*, 1373–1380.

Farret, A., Filhol, R., Linck, N., Manteghetti, M., Vignon, J., Gross, R., & Petit, P. (2006). P2Y receptor mediated modulation of insulin release by a novel generation of 2-substituted-5′-O-(1-boranotriphosphate)-adenosine analogues. *Pharmaceutical Research, 23*, 2665–2671.

Gertzen, C., Spomer, L., Smits, S., Häussinger, D., Keitel, V., & Gohlke, H. (2015). Mutational mapping of the transmembrane binding site of the G-protein coupled receptor TGR5 and binding mode prediction of TGR5 agonists. *European Journal of Medicinal Chemistry, 104*, 57–72.

Gremmel, T., Yanachkov, I., Yanachkova, M., Wright, G., Wider, J., Vishnu, V., Michelson, A., Frelinger, A. I., & Przyklenk, K. (2016). Synergistic inhibition of both P2Y1 and P2Y12 adenosine diphosphate receptors as novel approach to rapidly attenuate platelet-mediated thrombosis. *Arteriosclerosis Thrombosis and Vascular Biology, 36*, 501–509.

Guo, C., Chen, W., & Wang, Y. (2016). TGR5, not only a metabolic regulator. *Frontiers in Physiology, 7*, 646.

Hana, A., Deborah, M., & Ali, S. (2014). Secondary bile acids: An underrecognized cause of colon cancer. *World Journal of Surgery Oncology, 12*, 164.

Handelsman, Y. (2011). Role of bile acid sequestrants in the treatment of type 2 diabetes. *Diabetes Care, 34*, S244–S250.

Hauser, A., Attwood, M., Rask-Andersen, M., Schiöth, H., & Gloriam, D. (2017). Trends in GPCR drug discovery: New agents, targets and indications. *Nature Reviews Drug Discovery, 16*, 829–842.

Hauser, A., Chavali, S., Ikuo, M., Leonie, J., Martemyanov, K., Gloriam, D., & MadanBabu, M. (2018). Pharmacogenomics of GPCR drug targets. *Cell, 172*, 41–54.

Hill, M. (1990). Bile flow and colon cancer. *Mutation Research, 238*, 313–320.

Hochhauser, E., Cohen, R., Waldman, M., Maksin, A., Isak, A., Aravot, D., Jayasekara, P., Müller, C., Jacobson, K., & Shainberg, A. (2013). P2Y2 receptor agonist with enhanced stability protects the heart from ischemic damage in vitro and in vivo. *Purinergic Signalling, 9*, 633–642.

Hodge, R., & Nunez, D. (2016). Therapeutic potential of Takeda-G-protein-receptor-5 (TGR5) agonists. Hope or hype? *Diabetes Obese and metabolism, 18*, 439–443.

Ichikawa, R., Takayama, T., Yoneno, K., Kamada, N., Kitazume, M., Higuchi, H., Matsuoka, K., Watanabe, M., Itoh, H., Kanai, T., Hisamatsu, T., & Hibi, T. (2012). Bile acids induce monocyte differentiation toward interleukin-12 hypo-producing dendritic cells via a TGR5-dependent pathway. *Immunology, 136*, 153–162.

Jenkins, G., Cronin, J., Alhamdani, A., Rawat, N., D'souza, F., Thomas, T., Eltahir, Z., Griffiths, A., & Baxter, J. (2008). The bile acid deoxycholic acid has a non-linear dose response for DNA damage and possibly NF-kappaB activation in oesophageal cells, with a mechanism of action involving ROS. *Mutagenesis, 23*, 399–405.

Jiang, J., & Dingledine, R. (2013). Prostaglandin receptor EP2 in the crosshairs of anti-inflammation, anti-cancer, and neuroprotection. *Trends in Pharmacological Sciences, 34*, 413–423.

Kaia, M., Tarjei, H., Ellen, K., CR, K., & Lea, T. (2014). Activation of the bile acid receptor TGR5 enhances LPS-induced inflammatory responses in a human monocytic cell line. *Journal of Receptor and Signal Transduction, 35*, 402–409.

Kauffenstein, G., Tamareille, S., Prunier, F., Roy, C., Ayer, A., Toutain, B., Billaud, M., Isakson, B., Grimaud, L., Loufrani, L., Rousseau, P., Abraham, P., Procaccio, V., Monyer, H., de Wit, C., Boeynaems, J., Robaye, B., Kwak, B., & Henrion, D. (2016). Central role of P2Y6 UDP receptor in arteriolar myogenic tone. *Arteriosclerosis Thrombsis and Vascular Biology, 36*, 1598–1606.

Keitel, V., & Häussinger, D. (2011). TGR5 in the biliary tree. *Digestive Diseases, 29*, 45–47.

Keitel, V., Reinehr, R., Gatsios, P., Rupprecht, C., Görg, B., Selbach, O., Häussinger, D., & Kubitz, R. H. (2007). The G-protein coupled bile salt receptor TGR5 is expressed in liver sinusoidal endothelial cells. *Hepatology, 45*, 695–704.

Keitel, V., Spomer, L., Marin, J., Williams, R., Geenes, V., Kubitz, R., Haussinger, D., & Macias, R. (2013). Effect of maternal cholestasis on TGR5 expression in human and rat placenta at term. *Placenta, 34*, 810–816.

Khalid, S., Akram, U., Hassan, T., Nasim, A., & Jameel, A. (2017). Fully automated robust system to detect retinal edema, central serous Chorioretinopathy, and age related macular degeneration from optical coherence tomography images. *Biomedical Research International, 2017*, 7148245.

Kida, T., Tsubosaka, Y., Hori, M., Ozaki, H., & Murata, T. (2013). Bile acid receptor TGR5 agonism induces NO production and reduces monocyte adhesion in vascular endothelial cells. *Arteriosclerosis Thrombosis and Vascular Biology, 33*, 1663–1669.

Kim, L., Mertens, A., Maarten, R., & Hannah, M. (2017). Bile acid signaling pathways from the enterohepatic circulation to the central nervous system. *Frontiers in Neuroscience, 11*, 1–9.

Kumar, D. P., Senthilkumar, R., Sunila, M., Faridoddin, M., Grider, J., Murthy, K., & Sanyal, A. (2012). Activation of transmembrane bile acid receptor TGR5 stimulates both insulin gene transcription and insulin release in pancreatic b cells. *Biochemical and Biophysical Research Communication, 427*, 600–605.

Kwang-Hoon, S., Li, T., Owsley, E., Strom, S., & Chiang, J. (2009). Bile acids activate fibroblast growth factor 19 signaling in human hepatocytes to inhibit cholesterol 7α-hydroxylase gene expression. *Hepatology, 49*, 297–305.

Lane, J., May, L., Parton, R., Sexton, P., & Christopoulos, A. (2017). A kinetic view of GPCR allostery and biased agonism. *Nature: Chemical Biology, 13*, 929–937.

Lau, O., Samarawickrama, C., & Skalicky, S. (2014). P2Y2 receptor agonists for the treatment of dry eye disease: A review. *Clinical Ophthalmology, 8*, 327–334.

Lavoie, B., Balemba, O., Godfrey, C., Watson, C., Vassileva, G., Corvera, C., Nelson, M., & Mawe, G. (2010). Hydrophobic bile salts inhibit gallbladder smooth muscle function via stimulation of GPBAR1 receptors and activation of KATP channels. *Journal of Physiology, 588*, 3295–3305.

Li, T., Holmstrom, S. R., Kir, S., Umetani, M., Schmidt, D., Kliewer, S., & Mangelsdorf, D. (2011). The G protein-coupled bile acid receptor, TGR5, stimulates gallbladder filling. *Molecular Endocrinology, 25*, 1066–1071.

Lieu, T., Jayaweera, G., Zhao, P., Poole, D., Jensen, D., Grace, M., McIntyre, P., Bron, R., Wilson, Y., Krappitz, M., Haerteis, S., Korbmacher, C., Steinhoff, M., Nassini, R., Materazzi, S., Geppetti, P., Corvera, C., & Bunnett, N. (2014). The bile acid receptor TGR5 activates the TRPA1 channel to induce itch in mice. *Gastroenterology, 147*, 1417–1428.

Masyuk, A., Huang, B., Radtke, B., Gajdos, G., Splinter, P., Masyuk, T., Gradilone, S., & LaRusso, N. (2013). Ciliary subcellular localization of TGR5 determines the cholangiocyte functional response to bile acid signaling. *American Journal of Physiology: Gastrointestinal and Liver Physiology, 304*, G1013–G1024.

Michal, H., Talia, W., & Moshe, L. (2018). Bile acid receptors and the kidney. *Current Opinion in Nephrology and Hypertension, 27*, 56–62.

Min-Chan, C., Yi-Ling, C., Tzu-Wen, W., Hui-Ping, H., & Ming-Derg, L. (2016). Membrane bile acid receptor TGR5 predicts good prognosis in ampullary adenocarcinoma patients with hyperbilirubinemia. *Oncology Report, 36*, 1997–2008.

Muller, D., Zimmering, M., & Roehr, C. C. (2004). Should nifedipine be used to counter low blood sugar levels in children with persistent hyperinsulinaemic hypoglycaemia? *Archives of Disease in Childhood, 89*, 83–85.

Negus, S. (2006). Some implications of receptor theory for in vivo assessment of agonists, antagonists and inverse agonists. *Biochemical Pharmacology, 71*, 1663–1670.

Nishida, M., Sato, Y., Uemura, A., Narita, Y., Tozaki-Saitoh, H., Nakaya, M., Ide, T., Suzuki, K., Inoue, K., Nagao, T., & Kurose, H. (2008). P2Y6 receptor-Galpha12/13 signalling in cardiomyocytes triggers pressure overload-induced cardiac fibrosis. *EMBO Journal, 27*, 3104–3115.

Pathak, P., Liu, H., Boehme, S., Xie, C., Krausz, K., Gonzalez, F., & Chiang, J. (2017). Farnesoid X receptor induces Takeda G-protein receptor 5 cross-talk to regulate bile acid synthesis and hepatic metabolism. *Journal of Biological Chemistry, 292*, 11055–11069.

Perino, A., Pols, T., Nomura, M., Stein, S., Pellicciari, R., & Schoonjans, K. (2014). TGR5 reduces macrophage migration through mTOR induced C/EBPbeta differential translation. *Journal of Clinical Investigation, 124*, 5424–5436.

Peterson, T., Camden, J., Wang, Y., Seye, C., Wood, W., Sun, G., Erb, L., Petris, M., & Weisman, G. (2010). P2Y2 nucleotide receptor-mediated responses in brain cells. *Molecular Neurobiology, 41*, 356–366.

Phelan, J., Jerry Reen, F., Caparros-Martin, J., O'Connor, R., & O'Gara, F. (2017). Rethinking the bile acid/gut microbiome axis in cancer. *Oncotarget, 8*, 115376–115747.

Pols, T., Nomura, M., Harach, T., Lo Sasso, G., Oosterveer, M., Thomas, C., Rizzo, G., Gioiello, A., Adorini, L., Pellicciari, R., Auwerx, J., & Schoonjans, K. (2011). TGR5 activation inhibits atherosclerosis by reducing macrophage inflammation and lipid loading. *Cellular Metabolsim, 14*, 747–757.

Pols, T., Eggink, H., & Soeters, M. (2014). TGR5 ligands as potential therapeutics in inflammatory diseases. *International Journal of Interferon, Cytokine and Mediator Research, 6*, 27–38.

Rajagopal, S., Nalli, A. D., Kumar, D. P., Bhattacharya, S., Wenhui, H., Mahavadi, S., Grider, J. R., & Murthy, K. S. (2015). Cytokine-induced S-Nitrosylation of soluble guanylyl cyclase and expression of phosphodiesterase 1A contribute to dysfunction of longitudinal smooth muscle relaxation. *Journal of Pharmacology and Experimental Therapeutics, 352*, 509–518.

Ricardo, J., Rodrigues, A., Tomé, R., & Cunha, A. (2015). ATP as a multi-target danger signal in the brain. *Frontiers in Neuroscience, 9*, 148.

Rieg, T., Gerasimova, M., Boyer, J., Insel, P., & Vallon, V. (2011). $P2Y_2$ receptor activation decreases blood pressure and increases renal Na^+ excretion. *American Journal of Physiology: Regulatory Integrative and Comparative Physiolog, 301*, R510–R518.

Riegel, B., Lee, C., & Dickson, V. (2011). Self care in patients with chronic heart failure. *Nature Review: Cardiology, 8*, 644–654.

Sato, H., Antonio, M., Charles, T., Antimo, G., Mizuho, U., Hofmann, A., Régis, S., Schoonjans, K., Roberto, P., & Auwerx, J. (2008). Novel potent and selective bile acid derivatives as TGR5 agonists: Biological screening, structure-activity relationships, and molecular modeling studies. *Journal of Medical Chemistry, 51*, 1831–1841.

Schwiebert, E., Liang, L., Nai-Lin, C., Richards, W., Olteanu, D., Welty, E., & Zsembery, A. (2005). Extracellular zinc and ATP-gated P2X receptor calcium entry channels: New zinc receptors as physiological sensors and therapeutic targets. *Purinergic Signalling, 1*, 299–310.

Shreiner, A., Kao, J., & Young, V. (2015). The gut microbiome in health and in disease. *Current Opinion in Gateroenterology, 31*, 69–75.

Sil, P., Wicklum, H., Surell, H., & Rada, A. (2017). Macrophage-derived IL-1β enhances monosodium urate crystal-triggered NET formation. *Inflammation Research, 66*, 227–237.

Sriram, K., & Insel, P. (2018). G protein-coupled receptors as targets for approved drugs: How many targets and how many drugs? *Molecular Pharmacology, 93*, 251–258.

Thomas, C., Gioiello, A., Noriega, L., Strehle, A., Oury, J., Rizzo, G., Macchiarulo, A., Yamamoto, H., Mataki, C., Pruzanski, M., Pellicciari, R., Auwerx, J., & Schoonjans, K. (2009). TGR5-mediated bile acid sensing controls glucose homeostasis. *Cellular Metabolsim, 10*, 167–177.

Vauquelin, G., & Van Liefde, I. (2005). G protein-coupled receptors: A count of 1001 conformations. *Fundamental and Clinical Pharmacology, 19*, 45–56.

Wang, Y., Chen, W., Yu, D., Forman, B., & Huang, W. (2011). The g-protein coupled bile acid receptor, gpbarl (tgr5), negatively regulates hepati inflammatory response through antagonizing nuclear factor kappa light-chain enhancer of activated b cells (nf-kappab) in mice. *Hepatology, 54*, 1421–1432.

Wang, J., Liu, S., Nie, Y., Wu, B., Wu, Q., Song, M., Tang, M., Xiao, L., Xu, P., Tan, X., Zhang, L., Li, G., Liang, S., & Zhang, C. (2015). Activation of P2X7 receptors decreases the proliferation of murine luteal cells. *Reproductive Fertilization Development, 27*, 1262–1271.

Wang, L., Cheng, K., Li, Y., Niu, C., Cheng, J., & Niu, H. (2017). Glycyrrhizic acid increases glucagon like peptide-1 secretion via TGR5 activation in type 1-like diabetic rats. *Biomedicine Pharmacotheraphy, 95*, 599–604.

Watanabe, H., Nagesh, C., Laver, D., Seok Hwang, H., Davies, S., Roach, D., Duff, H., Roden, D., Wilde, A., & Knollmann, B. (2009). Flecainide prevents catecholaminergic polymorphic ventricular tachycardia in mice and humans. *Nature Medicine, 15*, 380–383.

Wenker, I., Sobrinho, C., Takakura, A., Mulkey, D., & Moreira, T. (2013). P2Y1 receptors expressed by C1 neurons determine peripheral chemoreceptor modulation of breathing, sympathetic activity, and blood pressure. *Hypertension, 62*, 263–273.

Whalen, E., Rajagopal, S., & Lefkowitz, R. (2011). Therapeutic potential of β -arrestin- and G protein-biased agonists. *Trends in Molecular Medicine, 17*, 126–139.

Wootten, D., Christopoulos, A., & Sexton, P. (2013). Emerging paradigms in GPCR allostery: implications for drug discovery. *Nature Review: Drug Discovery, 12*, 630–644.

Wu, L., Oshima, T., Fukui, H., Watari, J., & Miwa, I. (2017). Adenosine triphosphate induces P2Y2 activation and interleukin-8 release in human esophageal epithelial cells. *Journal of Gastroenterology and Hepatology, 32*, 1341–1347.

Yanguas-Casás, N., Barreda-Manso, M., Nieto-Sampedro, M., & Romero-Ramírez, L. (2017). TUDCA: An Agonist of the Bile Acid Receptor GPBAR1/TGR5 With Anti-Inflammatory Effects in Microglial Cells. *Journal of Cell Physiology, 232*, 2231–2245.

Yasuto, K., Kenjiro, S., Koji, N., & Tadashi, K. (2014). The Role of MicroRNAs in Ovarian Cancer. *Biomedicine Research International, 2014*, 249943.

Yui, S., Kanamoto, R., & Saeki, T. (2008). Deoxycholic acid can induce apoptosis in the human colon cancer cell line HCT116 in the absence of Bax. *Nutrition Cancer, 60*, 91–96.

Zetterberg, F., & Svensson, P. (2016). State of affairs: Design and structure-activity relationships of reversible P2Y12 receptor antagonists. *Bioorganic and Medicinal Chemistry Letters, 26*, 2739–2754.

MIX
Papier aus verantwortungsvollen Quellen
Paper from responsible sources
FSC® C105338

If you have any concerns about our products,
you can contact us on
ProductSafety@springernature.com

In case Publisher is established outside the EU,
the EU authorized representative is:
Springer Nature Customer Service Center GmbH
Europaplatz 3, 69115 Heidelberg, Germany

Printed by Libri Plureos GmbH
in Hamburg, Germany